Under
the Sky
We
Make

Under the Sky We Make

How to Be Human in a Warming World

Kimberly Nicholas, PhD

G. P. Putnam's Sons

New York

PUTNAM
— EST. 1838 —
G. P. Putnam's Sons
Publishers Since 1838
An imprint of Penguin Random House LLC
penguinrandomhouse.com

E-mail on page 83 is reprinted courtesy of Jake Milner.
E-mail on page 91 is reprinted courtesy of Elena Deegan-Krause.
E-mail on page 244 is reprinted courtesy of Colt Tipton-Johnson.

LIBRARY OF CONGRESS CATALOGING-IN-PUBLICATION DATA
has been applied for.

ISBN 9780593328170 (hardcover)
ISBN 9780593328187 (ebook)

Printed in the United States of America
1   3   5   7   9   10   8   6   4   2

Book design by Nancy Resnick

To 2030. I hope we did right by you.

Contents

Contents

Under the Sky We Make

Introduction

Science Won't Save Us

I n the half lifetime since we met in college, I've had many adventures with my friend Colty. Together we've seen the sun rise through towering pines after dancing all night on the shores of Fallen Leaf Lake; buoyed each other through heartbreak and celebrated when we both found true love; sobbed through a funeral where our tears loosened the glue of the fake mustaches we wore to honor our dead friend's iconoclastic spirit.

Colty and I have always been able to talk about anything and everything, but I was surprised by the conversation we had a few years ago, when I called to congratulate him on the birth of his second child. I was expecting to hear intimate stories of adjusting to new family routines, but his daughter's arrival had raised Colty's sights much higher.

"Kimmy, I want to talk about climate change," he said. "I feel like all these red flags are going up around me, saying, 'Wake up!' But I'm overwhelmed. Where do I start?"

For decades, I've spent my professional life researching, teaching, and communicating about climate change. But until recently, my climate work hasn't been something I talked about with my closest friends. For my research, I spent long, sweaty days of field-work in the mountains and grasslands and vineyards of California, documenting sagebrush marching uphill and Pinot Noir grapes around my hometown of Sonoma losing their color as the world warmed around them. All this felt like a separate world from the time I spent with my friends, where we talked and joked about our families, careers, and love lives; went for hikes through the red-woods and to the beach; and shared good meals and wine.

Through my years of avoiding talking about the climate ele-phant in the room, I was like most Americans, who know climate change is happening and are worried about it. (Only 10 percent don't believe the unequivocal fact that humans are warming the climate.) But still, unlike the vocal climate dismissives, most of the climate-concerned majority stays silent, reporting they almost never talk about climate change with friends and family. Many feel like it's someone else's problem: polar bears, perhaps, or politi-cians, or people unlucky enough to be born somewhere poorer or sometime later than they were.

Once you do start talking about it, you can't avoid scientific truths with overwhelming, existential implications. Human cli-mate pollution and destruction of nature are putting at risk both human civilization and life on Earth as we know it. Gulp. In the face of this enormity, no wonder so many feel helpless (that you're powerless and don't matter) and hopeless (that no one can help and therefore nothing matters). I understand those feelings. I have them myself.

Over the course of my career, climate change has transmogri-fied from something only experts could see—reading clues trapped in icy air bubbles or statistical patterns in long-term data sets—to something everyone on Earth is living through. For me, climate change has gone from being something I study to a way that I see the world and experience my life. It's one thing to publish a study on the hypothetical impacts of temperature increase on California's people and ecosystems; it's another to feel my stomach gripped by fear as my parents flee a catastrophic California wildfire cranked up by longer, hotter, drier summers. It's one thing to measure declining color pigments in Pinot Noir grapes due to increasing temperatures; it's another to viscerally mourn the loss of the taste of my favorite wine as it passes from this Earth.

By the time Colty and I finally talked about climate, it had gone from a measurable change to a prevailing crisis to a screaming emergency. Climate change was already rewriting the stories we read, reshaping our everyday lives, affecting everything and everyone we love. Climate was already woven through everything else we talked and cared about.

Colty and I had a series of climate counseling sessions, where we talked about the core values he hoped to pass on to his kids, his hopes and fears for the future, what inspired him and kept him going. In our conversations, I translated the science of an enormous global problem to a personal and human scale. I wanted to help my friend see the power he *already* had to be a force for climate and social good and to live a better life more in line with his values along the way.

These sessions were a version of the conversation I've been having over and over again, for years—with strangers at parties and on trains, in my talks with business leaders, festivalgoers, protesters, knitting grandmas, and everyone else who will listen. From these conversations, I know firsthand that there are many smart, concerned people out there who are deeply invested in the future of humankind and who don't need further proof that climate change is a real and urgent threat. Those of us who want to help are the majority; if even a fraction of us can mobilize and take action, we are more than enough to stabilize the climate.

That's why I'm writing this book.

Science Won't Save Us

Here's some good news for everyone who's been avoiding science since you had to memorize the periodic table in high school: We don't really need more science to solve the climate crisis. Saving the planet for humanity (and the rest of nature) is no longer a matter of understanding anything we don't, or developing a ton of technology we haven't. Science has carried our collective knowledge about as far as it can in the time we have. Luckily, it's enough.

The science of climate change is firmly settled, and has been for a very long time. It boils down to just five key facts, which I've been teaching since 2011: It's warming. It's us. We're sure. It's bad. But! We can fix it: Humans have the capacity to stop dangerously destabilizing the climate.

Basically, the climate problem has been solved on paper many times over by now. We know what we have to do and how to do it. Further tech breakthroughs could make it even faster and cheaper, but essentially we have the tech we need in hand.

Okay, then fixing climate change is a political problem, right? Well, yes and no. Changes in goals, policies, and laws are crucial. And you need to understand some power politics to get why humans find ourselves threatening to unplug our own life-support system in the first place. But the basic political framework to solve the climate crisis is also in place. To paraphrase three decades of negotiation-speak: The world has agreed that stopping climate warming is in the shared interest of humanity. The 2015 Paris Agreement says each and every country (and therefore, each and every industry and city, and ultimately each person) must do their fair share to stop climate heating before it exceeds intolerable limits. That bounces the ball right back to us.

So yes, we need technical and social transformations, informed by science, to be put into practice to solve the climate crisis. We need political processes to do this fairly, citizens informed by fact-based climate education at all levels, and media coverage that explains how climate shapes nearly every story. Preferably all at once.

But what will really make or break the climate, globally and forever, is what ordinary people do in the next decade. The climate hinges on what people vote for, with our ballots and time and money and careers, with our leisure and travel and consumption and production, with our relationships and conversations and aspirations and memes and everything else that adds up to culture.

Science or experts or technology isn't enough to save us from climate catastrophe. We as humanity, a groundswell of people alive today around the world, have to save ourselves, through what we think and feel and ultimately what we do. This means we need people with the courage and compassion and imagination to transform themselves, and society, in the ways that science tells us are necessary to maintain conditions for life on Earth to be able to thrive. Each of us can become that sort of person; more and more are every day.

The climate especially hangs on what Americans do in the next decade. In 2010 I moved from California to Sweden, where I'm a professor of sustainability science at Lund University. Looking at the United States from abroad, it's acutely clear how much the climate hinges on what happens in the world's largest economy and largest historical climate polluter (USA! We're number one!). For practical reasons, Americans starting to take climate responsibility at scale would be an enormously powerful accelerator of climate action worldwide.

But there's a deeper cultural element too: For generations, the world has aspired to an American model of consumption that is widening inequality, making us sick and unhappy, and destroying the workings and wonders of the natural world. Americans need to reinvent our dream toward one worth striving for and spreading: a good life for *all*, within a budget the planet and future generations can afford.

That means the climate *really* needs people like me and Colty to step up, because our privilege translates into disproportionate power and responsibility in many realms—and the climate is no exception. With our incomes in the top 10 percent globally, we're part of the group that consumes the majority of the world's resources and therefore creates most of the world's problems when it comes to heating the climate and destroying nature. (If your income is more than $38,000 per year, you're in the top 10 percent of the global income camp too.) Those of us with even more climate privilege, who are higher up the income ladder, are having an even more outsize impact. We need to take a long, hard look at our climate legacy—in terms of both our personal lifestyles and the political, economic, and cultural systems we help create, support, and empower—to see if it's really the mark we want to leave on the world.

It's Us

This book is a guide to how to live through what I think is the most riveting, challenging, terrifying, important, and meaningful time to be alive. The most essential lesson I take away from the thousands of scientific studies I've read, and the fifty-plus I've written, is that people alive right now are living through the decade that will define the future for both humanity and life on Earth. We are setting the global thermostat and therefore the boundaries and possibilities for human development, as well as the living conditions for all life on this planet, for the rest of our lives . . . and affecting millennia to come.

The goal is for humanity as a whole, and I hope you and me personally, to get through the warming decades ahead together; to live to see a time where the climate is stabilizing, nature is recovering, human well-being is flourishing, and equality is increasing. To get there, we will have to draw on the best of ourselves and bring out the best in one another.

How can we stop climate breakdown in time to protect the only home we have, which we share with almost 8 billion fellow humans and about 8 million other species? In my opinion, we need to put the human values we hold most dear at the core of this work, because ultimately preserving the legacy of these human values is what is at stake. Nothing short of transformative change is going to be enough.

As you may have guessed by now, this is a different kind of climate book, one that's not only about the science.

Sure, I draw from my expertise as a scientist, with a reverence for evidence (hence all those endnotes) and analytical rigor. But my mission is to use this science to draw lessons about how to be human in our warming world. I aim to give the science a human face, by sharing stories from my own journey and from people who inspire me. By sharing these stories, I am acknowledging something that was initially hard for this scientist to admit: It is not just facts that we need to solve climate change; it is also tapping into the strength of our feelings about what we most fear, grieve, and love. It is the people and places and things I love—relationships, family folklore, beloved landscapes, wine—that motivate and sustain my work. Only by looking at climate change as humans, bringing all our humanity and empathy to bear, can we start to head toward the solutions at the speed and the scale needed.

Breaking Up with Exploitation

As a nerd steeped in sophisticated analyses, when I look for the root cause of humanity's current woes, I'm embarrassed by the naivety of my diagnosis. But here it is anyway:

Right now, too many people have a mindset of exploitation of nature and other people. This Exploitation Mindset is not fact. It's not based on any fundamental truths about nature or human nature. Instead, it's a story we're telling ourselves about who we are and the way we live in middle-class, industrialized societies, until it's become as invisible and taken for granted as the air we breathe.

This story is changeable, even though it feels like it isn't. There's a better way to live. We must find it.

Our task as humans in this warming decade and beyond is taking the science and using it as a lens to change not just our systems but also ourselves, from the inside out. By clarifying our values and shifting our mindsets and actions, we can start to change the world.

We need a new mindset to have a good future for life on Earth. We can and must change the story of exploitation, by identifying all the ways this mindset is baked into our current society, eradicating it, and replacing it with a better one.

My suggestion for a better story is what I call the Regeneration Mindset, which is focused on working *with* rather than *against* nature and bringing out the best of ourselves and one another. I've boiled down the Regeneration Mindset to three ideas that could be embroidered on your grandma's throw pillow or pasted in construction paper on your kindergartener's classroom wall: Respect life. Stop harming life. Strengthen life. Sounds pretty basic, but as I hope to illustrate in this book, I think these principles are flexible enough to guide action across diverse circumstances. Actually putting them into practice would be profoundly transformative.

The Path Ahead

This book is structured in three parts, which roughly correspond to the brain, the heart, and the hands, or thinking, feeling, and doing. In general, the book progresses from diagnosing problems to offering solutions, and from the global and abstract to the personal and tangible.

In Part 1, I tell stories of my family history and the history of life on Earth, to illustrate how humans and nature are tied up together (it's warming; it's us) and show the legacy humans are leaving in the sky and across the living world. I tell these stories because I believe understanding the science—truly understanding what the workings of the material world we share mean for humanity and our civilization, and how acute our predicament is right now—offers a kind of awakening. I hope coming to terms with the sobering power humankind wields at this moment helps us converge on the urgent need to lay down the weapons of the Exploitation Mindset and pick up the tools of Regeneration instead. The point of Part 1 is to empower you to orient your goals by asking the questions and having the conversations about how your everyday life and everything you love are inextricably connected to the climate and the living world, which are under urgent threat.

Well, *that* was all a bit heavy, now, wasn't it?! If Part 1 is about facts, Part 2 is about feelings, drawn from my own journey of slowly learning to acknowledge all the uncomfortable emotions of being a climate expert with a brain, and a human being with a heart, in a warming world. I share what losing a dear friend to cancer taught me about grieving climate losses (it's bad), how facing my climate fears with my community gives me strength to carry on, and my enraging experiences of being a scientist in a world that sometimes doesn't want to hear the truth (we're sure). To see a pathway out of our climate and ecological crises, I had to stop looking to science for all the answers and start changing myself, using the climate crisis as a crucible to create meaning by clarifying my values and putting them into practice. The goal of Part 2 is to help you find your climate calling: to identify and nurture what really matters to you and to cultivate and strengthen the personal and community resilience essential to make your way in this warming world with kindness and purpose.

After we've made it through All the Climate Feels in Part 2, we're ready to roll up our sleeves and get to work. Part 3 is all about how We Can Fix It: envisioning and creating a future we want through both personal and system change. Here's where "climate action" goes from a hollow hashtag to concrete steps for who can do what to zero out climate pollution while strengthening people and nature. This part is about putting the Regeneration Mindset into practice within the day-to-day life you already lead, and how it scales to policies. It's also about expanding your vision of your sphere of power, to help extend the reach of what humans make possible. In line with my focus on what matters most, I emphasize what research and my personal experience show are the most effective ways to spend your limited time and energy for maximum impact. The goal of Part 3 is to help you find ways to use your unique gifts with agency, urgency, and joy to start bending our story from a legacy of harm to one of care.

Simply put: This is a climate scientist's book about the apocalyptic urgency of prioritizing not just the planet but also our humanity. I want to tell the story of our Earth's past, our world's present, and humankind's future—under the sky we make.

Part I

It's Warming. It's Us.

How We Got Here

Chapter 1

Carbon Is Forever

Understanding the Urgency of the Task Ahead

M y mother's mother's mother, Clara, fled what is now Ukraine in 1904, when she was twenty-two. She had sewn her fili-greed platinum engagement ring into her jacket to avoid detection as a deserter. If the authorities caught you leaving with your husband, they knew you were escaping for good. Her immigration record from Ellis Island lists her port of departure as Bremen. She and her husband, my great-grandfather Mark, lived in a damp tenement near Coney Island before they eventually settled in Denver, where they ran a women's clothing shop and raised my grandmother Lillian and her brother.

I've seen only one photograph of Mark, wearing a fedora, and Clara, with dark wavy hair. It was taken on a suburban Denver street with my mother, a serious five-year-old, and her sister Judy, already a great beauty at nine: old-world grandparents who loved borscht, posing with their wholly American grandchildren who thought the smell of beets and cabbage cooking was just awful. Clara made her life in a new country in her twenties, as I did in my thirties when I crossed the ocean to live in Sweden.

I never met Clara, but she touches my daily life in two ways.

First, her diamond sparkles on my left ring finger. Second, carbon from the coal that powered her escape, across first a continent and then an ocean, is still warming the atmosphere I share today with nearly 8 billion people. Because when your individual actions are powered by fossil fuels, some of the carbon from those actions stays in the air for thousands of years. Your story doesn't end with your death; its contrails unfurl in the physical world for millennia.

Clara lived to be eighty-two—a good, long life. Her grandchildren—my mother and her two siblings—are the last generation of my family to have known her. They're now grandparents themselves. Once they're gone, living memory of Clara will wane and eventually the stories they shared of her will disappear too. Clara's life, as real and as vivid and important as mine or anyone else's, will fade into the background of the human tapestry. But her carbon will outlast us all.

I don't know the name of Clara's mother's mother's mother. She would have been born in black-soil country sometime around 1800, so I can guess that she was part of a big family, all of whom worked hard on the farm. I like to imagine them playing music around the fire at night. But here's one thing I know for sure: A portion of the carbon sent skyward from the wood they burned to stay warm—and the carbon they released plowing the rich black soil—is still in our air today, and it will be for at least the next *three hundred* generations.

I don't know what Clara was thinking when she decided to risk the perilous journey to a new land and leave behind everything she knew. I don't know how much thought she gave to her potential descendants and the life they would have as a result of her choice, or how much she was motivated by her own more immediate desires. Nevertheless, she set in motion a chain of events that shaped my life, giving me more choices, more freedoms, more privilege. I'm deeply grateful to her as a good ancestor.

Everyone alive today is skywriting the most important legacy of their lives in atmospheric carbon. Long after our names and faces and deeds have faded from living memory, long after any genetic or creative or physical or digital traces of us are gone, this carbon legacy will define us in the minds and stories of our distant descendants. It will literally define the terms of their lives: where they can live, how they can make a living, what kind of civilization and nature surround them. We will be remembered for our carbon legacy by far more people than we'll ever share a meal with or know by name.

Carbon Is Forever

"A diamond is forever." That De Beers cliché, in use since 1947, is to a certain extent true. Diamonds are composed of pure carbon. Carbon is the building block of life, found in all known life-forms. Living bodies, from plants to humans, are first and foremost water, which helps regulate temperature, circulate nutrients, and flush waste. But this temporary and variable internal lake drains away to join a new river when we die. What remains is largely carbon. The same carbon atoms have been circulating in a marvelous cycle between air, rocks, soil, water, and living creatures for about half a billion years, since land plants figured out how to turn sunlight and air into humble mosses, then towering trees, greening the earth with the magic of photosynthesis.

How much carbon is in which stage of the cycle largely controls the climate and therefore the habitability of planet Earth. Climate is the long-term average of weather in a particular place. As the saying goes, climate is what you expect; weather is what you get. Carbon in the atmosphere acts like a thermostat. It is the atmospheric level of carbon that primarily determines the temperature to which the planet will eventually heat. When humans burn plants, we put the carbon those plants drew from the air to build their leaves and trunks back in the atmosphere. In the chemical reaction of burning, two oxygen atoms glom on to one carbon atom, forming carbon dioxide. CO_2 represents 75 percent of heat-trapping emissions; methane and nitrous oxide make up most of the rest.[*] These greenhouse gases trap heat near the Earth's surface instead of letting it escape into space. This extra heat warms the air and land and upper oceans.

Looking back over time, across the world, continents and oceans sing a coherent chorus: It's warming! The evidence for warming is found in decades of satellite and ocean buoy records; centuries of land thermometer and written records, like Thoreau's Concord journal, from which we know that plants now flower earlier in warmer springs than they did in his time; and millennia of environmental records quietly curated in the bodies of trees and corals and in the pollen lining lake beds. From these ancient records, we know the climate has been relatively stable for the last ten thousand years. Not coincidentally, human civilization was founded and flourished during this benign climate.

Humans have now markedly warmed the world, by approximately 1°C[†] above its pre-industrial average. Compared with his-

---

[*] A note on nomenclature for the sticklers: Throughout this book, I use "carbon" to refer to carbon dioxide emissions. "Climate pollution" or "greenhouse gas emissions" refer to the sum of all human emissions warming the climate. Note that "carbon footprint" includes all greenhouse gas emissions, expressed in carbon dioxide equivalents.

[†] A note for American readers: All temperatures in this book are given in degrees Celsius, a very sensible temperature scale where water freezes at 0°C and boils at 100°C. All scientific studies and every country in the world save five (the Bahamas, Belize, the Cayman Islands, Liberia, and the United States) use Celsius to measure temperature.

Every mention of temperature in this book is about relative temperature, comparing the difference between temperatures (generally an increase from the pre-industrial global average temperature to a current or projected future temperature).

tory, today's warming is shockingly fast. A new color scheme, purple, had to be introduced in maps of Australia and the globe to show how much warming has *already* occurred above the previous end of the scale, red.

We know that today's warming is caused by humans through a combination of observations, theory, and models. Observed measurements show that CO_2 has increased 40 percent in the air; enough carbonic acid has been formed from CO_2 dissolving in the oceans to increase their acidity 26 percent. Atmospheric measurements sensitive enough to distinguish tiny variations in the global concentration of CO_2 pinpoint the source of this carbon from cities and factories. The warming observed is greater over land than over oceans, at higher latitudes, in the upper oceans, and lower in the atmosphere: all consistent with what physics predicts from humans burning carbon on land. Humans have likely caused *more than 100 percent* of warming to date, because natural trends alone (from variations in the sun and volcanic outputs) would have caused a slight cooling. In short: It's us.

For millennia, essentially the only carbon humans added to the atmosphere came from cutting and sometimes burning plants that grew within their lifetimes ("green" carbon), and from plowing and cultivating soil (a huge library of organic carbon built by generation after generation of decomposed plant roots and leaves). About 30 percent of the cumulative carbon humans have emitted to the atmosphere comes from centuries of transformation of land, clearing vegetation and disturbing soil to raise livestock, grow crops, and harvest timber. Human exploitation of land still causes about a quarter of our total climate pollution, including most methane and nitrous oxide emissions, as I'll cover in Chapter 10.

But the lion's share of greenhouse gas emissions today,

Each 1°C temperature change is equal to 1.8°F. So a temperature increase of 1.5°C is 2.7°F warmer; an increase of 4°C is 7.2°F warmer. Okay, no more footnotes!

including 86 percent of the carbon emitted over the last decade, is created by our burning of *ancient* plants. This fossil carbon is the 300-million-year-old remnants of lush swamps and sludgy seabeds (*not* dinosaurs!), transformed by heat and pressure from the Earth into fossil fuels like coal, oil, and gas. When we burn coal and gas to make electricity and energy to power and heat our homes, and oil to move our cars and planes, that long-buried carbon is transferred from the Earth to the atmosphere. It carries a telltale isotopic signature that definitively identifies it as fossil carbon. And we currently burn a metric shit-ton of prehistoric carbon: 36.8 billion tons of fossil CO_2 in 2019, to be exact.

Currently, nature cleans up a bit over half of humanity's carbon mess for free. Slightly more than a quarter of our carbon dioxide becomes plant food for photosynthesis on land. The oceans take up a bit less than a quarter of our carbon, but with a consequence much worse than more trees. Ocean acidification opposes the chemical reaction that the tiny creatures at the foundation of marine food webs use to build their shells out of basic calcium carbonate. The science writer Elizabeth Kolbert likens it to "trying to build a house when someone keeps stealing your bricks." There is a threshold beyond which it is physically impossible for calcium-based shells to form. Reaching it would be disastrous for life in the oceans.

After the land and oceans absorb their share, much of humanity's carbon pollution remains in the atmosphere, driving warming essentially forever. Up to a quarter of the carbon released into the atmosphere today will remain there *ten thousand years later,* a birthday Stonehenge is slightly more than halfway to reaching, and the Great Pyramids slightly less. Its route out of the atmosphere happens at the unimaginably subtle rate at which raindrops flatten granite mountains, or at which tectonic plates cross oceans, moving about as fast as your fingernails grow. Over millions of years, most of this carbon will end up at the bottom of the ocean, some to eventually be returned to the sky when carbon-rich seafloors push under continental plates, melt to magma, and release their carbon back to the atmosphere when they erupt out of volcanoes.

Before humans showed up, the carbon slowly leaking from rocks to sky via the oceans and volcanoes was balanced by the carbon dissolving from sky to rocks in rainwater. Today, our factories, cars, and other industrial activities emit more carbon in three days than all volcanoes do in a year.

Humans are now adding carbon to the atmosphere hundreds of thousands of times faster than geology can remove it, hence our skyrocketing concentration of airborne carbon. The concentration was around 280 parts per million (ppm) back in the pre-industrial days. It passed 300 ppm for the first time in at least eight hundred thousand years in the early twentieth century, then started rising rapidly after 1950. In 2019, it averaged 411 ppm, rising about 2.3 ppm per year.

The "safe" level is 350 ppm, which we blew past in 1987. Oops.

All this carbon buildup has consequences: Every kilogram added to the atmosphere pushes us toward more dangerous climate change. The carbon math is brutally simple: the more emissions, the more warming, the more harm and suffering.

The Science and Politics of Danger

While the sluggish politics have always lagged far behind the urgent science, stopping warming in time to avoid dangerous climate change has been the central purpose of international climate negotiations for three decades. The objective of the world's first international climate agreement, the 1992 United Nations Framework Convention on Climate Change, and all subsequent ones, including the Paris Agreement, is to achieve "stabilization of greenhouse gas concentrations in the atmosphere at a level that would prevent dangerous anthropogenic [human-caused] interference with the climate system." Danger is defined in terms of three critical priorities that must be protected: healthy ecosystems, food production, and sustainable development. Seems reasonable enough. But of course, nothing is ever easy when it comes to international politics.

A long political process established the temperature goals meant to achieve the purpose of climate stabilization in time to avoid dangerous climate change. In the historic 2015 Paris Agreement, the nations of the world agreed to hold the increase in global average temperatures "well below 2°C above pre-industrial levels . . . pursuing efforts to limit the temperature increase to 1.5°C."

The definition of dangerous climate change enshrined in international agreements boils down to a simple affirmation that humans should maintain a climate compatible with life and prosperity. Avoiding dangerous climate change means we stabilize the climate in time to "allow ecosystems to adapt naturally to climate change, to ensure that food production is not threatened, and to enable economic development to proceed in a sustainable manner." These sound like pretty darn good ideas. I want to live on a planet where we maintain conditions where nature can survive, humans can grow food, and all people have the opportunity for a good life.

Unfortunately, humans have *already* caused warming of about 1°C, and this warming is *already* undermining each of the three priorities that define a safe climate. I'm sorry to tell you that this means we already live in a world of dangerous climate change.

First, the climate is already changing faster than ecosystems can naturally adapt. The current human-caused increase in CO_2 is more than *one hundred times* faster than the natural variation that happened in the past. Picture strolling along a NASCAR racetrack and trying to keep up with the cars zooming by; keeping up with climate is an even bigger challenge. Some species are running out of options. Gorgeous birds in Hawai'i, including honeycreepers that exist only there, live in the remaining high-elevation forests, where cooler temperatures protect them from mosquitoes and the malaria they carry. As warming brings the shrill whine of the mosquitoes steadily higher, the birds are getting squeezed off the mountaintops toward extinction. Scientists are making desperate attempts at "assisted colonization" to relocate some species and create an "insurance population" to buy more time, but no one thinks this approach will save unraveling ecosystems. Camille Parmesan, who has studied ecosystems under climate change for decades, has found that half of all species studied have had to move in space (range shifts) and two-thirds have moved in time (for example, flowers blooming and birds migrating earlier in the spring) to chase after their niche in a rapidly warming world. These large responses to "only" 1°C so far are deeply worrying for what's in store under more warming.

As for the second component of danger, climate change is already threatening food production. According to a 2019 study led by Deepak Ray, climate change has been stealing 1 percent of consumable calories across ten major staple crops since 1974 (losing enough food to feed 50 million people each year). Earlier work by David Lobell and colleagues looked at decades of yields for four crops that provide 75 percent of human calories, finding they declined in warmer years at a rate of about 10 percent for each 1°C of warming experienced. Land degradation is further stressing food production; it has already reduced crop productivity in almost a quarter of global land area.

And climate change is already slowing economic development and increasing inequalities, undercutting human progress and the physical security on which it rests. A global analysis by Stanford

colleagues showed that human-caused warming has already increased the income gap between rich and poor countries by 25
percent, disproportionately slowing economic growth in poorer
countries, which tend to be warmer.

Focusing on a broader set of impacts beyond this UN definition of danger, a review of 3,280 research papers on current
impacts concluded that "greenhouse gas emissions pose a broad
threat to humanity," and the few nonharmful impacts they found
could not counterbalance any of the harms "related to the loss of
human lives, basic supplies such as food and water, and undesired
states for human welfare such as access to jobs, revenue and security." Climate change is already bad, and it's poised to get worse.

The Carbon Bathtub Is Almost Full

Climate action needs to be massively accelerated to meet the
Paris temperature goals, because current climate policies are "blatantly inadequate" to do so (more on that shortly). To understand
what humanity has to do to stabilize the climate, imagine for a
moment that our atmosphere is like a bathtub, one in which you
are a happily floating rubber ducky. The volume of the bathtub is
its capacity to store a set amount of water. The water level represents the concentration of carbon in the atmosphere. From the
faucet pours the carbon humans burn and unlock from ecosystems, raising the level. In my lifetime so far, we have kept opening
the tap wider, transferring ever more carbon from Earth to sky,
well after the science of the dangers of climate change was clearly
established.

Out the drain flows the half of our carbon pollution that nature
is currently absorbing on land and at sea. (Worryingly, warming
is constricting the size of the drain, because overheated and over-

used forests, stressed by more pests and fires, cannot take up as much carbon. A warmer ocean is like a can of Coke left open in the blazing sun; it quickly loses its carbonated fizz.)

But for now, this leaves about half the water that comes out the tap each year remaining in the bathtub. Thus, the water level is rising. You used to splash about with rubber duckies in the safe, cozy confines of the tub, with an infinite and distant sky beyond. But over your lifetime, your view has been steadily changing. As the water level rises and carries you with it, you see the edge of the tub drawing ever closer, and beyond that, there is an unknown world.

The edge of the tub is the amount of carbon we can burn and still stay within a given amount of warming, say, 1.5°C. This is our finite, not-very-large *carbon budget.*

Alarmingly, the carbon bathtub is already nearly full. We are perilously close to the edge of the bathtub today. There are methodological nuances in how best to count the carbon budget, and of course the remaining budget is bigger if humanity is aiming for the less stringent 2°C rather than the 1.5°C goal. However, there is clear scientific consensus that the budget is limited, that something like 90 percent of it has already been used up, and that we are quickly exhausting what little budget remains.

When my great-grandmother Clara was born, the thousands of human generations before her had collectively used up essentially none of the carbon budget. When my grandparents were born, about 95 percent still remained. How far away that bathtub edge must have looked from the bottom of the tub! For my parents, it was nearly the same, nearly 90 percent left, a seemingly infinite sky. Their generation used up a lot more of the carbon budget, leaving less than 70 percent remaining for me and my generation when I was born in 1978.

In the four decades since the first *Star Wars* movie was in theaters, humans have used up almost 60 percent of the carbon budget available for all time, for all of humanity. If we continue burning fossil fuels at today's rate, before 2030, before my *Frozen*-loving fairy goddaughter graduates from high school, it will be entirely gone, depleted, overspent. A priceless resource that should be stewarded and handed down across all generations of the human family has literally gone up in smoke, in the blink of a geological eye, with its benefits very unequally distributed.

Because carbon is essentially forever, the carbon budget is forever too. If I use up more than my share, this leaves less space for you. This is true today across places: between rich and poor countries and between high- and low-emitting individuals. This tug-of-war is also true stretching across time: between previous generations, those of us alive today, and all humanity to follow.

The key insight from the carbon bathtub is that, to stabilize the climate, carbon emissions have to be not just slowed or even slashed. Because some carbon lasts forever and the bathtub is about to overflow, carbon emissions have to fall *all the way to zero. Any* carbon added to the atmosphere will prevent stabilizing the climate. (Emissions of other greenhouse gases with a much shorter atmospheric lifetime must be strongly reduced from today's levels, but not all the way to absolute zero.) Warming will only stop after "net zero" carbon is reached; this is the point where our carbon emissions from the bathtub tap match our removals from the drain, so the water level is no longer rising. In practice, the potential to increase carbon removals by "widening the drain" is very limited. This means nearly every country, industry, sector, city, and individual must *completely stop* adding carbon to the atmosphere to achieve real, absolute zero emissions in the first place.

Any carbon you still add to the atmosphere means either someone else (likely someone poor, young, or both) has to cut their emissions more to clean up your mess, or you have to physically remove carbon from the atmosphere. While the Exploitation Mindset favors a range of "pollute now, pay later" schemes to keep burning carbon, carbon removal is a ginormous, expensive, and risky undertaking. There is low confidence that it will actually work. The definitive climate authority is the Intergovernmental Panel on Climate Change (IPCC), a huge team of scientists who synthesize the latest evidence in massive reports after exhaustive scientific and governmental review. The latest IPCC assessment warns that carbon removal schemes would need to be deployed at large scale for at least one hundred years to have a two out of three chance to "significantly reduce" CO_2 concentrations in the atmosphere. While some carbon removal may be necessary to reach net zero, *in addition* to stopping emissions, clearly it's not Plan A—rapidly stopping fossil fuel use is!

What's Ahead?

Yikes, we already live in a world made more dangerous by human-caused warming, and to stop warming we have to completely stop adding carbon to the atmosphere.

Until we do, what's in store for the world we know over the twenty-first century?

To help orient your thinking to the scale of a century, please do some quick morbid mental math to estimate how much more of this century you and your loved ones can expect to live through. To help you calculate: Life expectancy more than doubled in the twentieth century and is now over seventy globally and over eighty in industrialized countries. (Personally, I expect to retire around 2050 and live into the 2060s; if I live as long as my father's grandmother, I will die in 2079.) Even if you don't count on having many more years on this planet, please keep in mind that more than 40 percent of the world's population in 2020 is under the age of twenty-five, hopefully including many people you love; these young people have a good chance of living to see the year 2100.

Do you have in mind where you and your loved ones will be in 2030, 2050, and beyond? Good, let's look at the warming that could happen this century.

How much warming our planet experiences depends on the total, cumulative amount of carbon in the atmosphere. The faster we stop adding more carbon, the sooner the planet's temperature will stabilize.

Looking at the coming decade, I've got bad news for the instant-gratification wired among us. Some further warming, and thus worse impacts, is unavoidable at this point, already baked into the system; it is going to get worse before it gets better. That's because the next few decades are the era of committed climate change, where the global climate system will keep warming as long as humans are adding any carbon to the atmosphere. We will not start to see climate stabilization for decades, even if we throw ourselves at this problem with the urgency it demands and zero out emissions quickly, because each and every car and factory and power plant still belching out carbon will cause added warming. There will be additional, unnecessary human suffering and serious ecosystem destruction, compared to a world where we had stopped emitting carbon earlier. Even at "only" 1.5°C of warming, and even taking all available adaptation measures, the IPCC warns that extreme events will be more frequent and deadlier, air will be harder to breathe, diseases will spread faster and wider, and mental health and violence will be worse.

Some inevitable warming is certainly bad news, but that doesn't mean we should give up. To the contrary, our actions to cut emissions now are critical. How much *more* warming our planet will experience beyond 1.5°C depends mostly on how much carbon humans keep adding to the atmosphere; that is, what we do now, and soon.

Looking further out, let's consider a possible range of warming from 1.5°C to 4°C. To put these numbers in context, it's helpful to know that an Ice Age is about 4°C colder than pre-industrial temperatures. A 4°C temperature change produces a different planet than the one we grew up on. During the last Ice Age, ice covered a third of the planet (instead of a tenth today). Boston was buried under nearly a mile of ice.

Consider that the planet's temperature is usually as stable as the human body's. As climate scientist Katharine Hayhoe points out, a body temperature one degree above normal is a fever; with a fever of two degrees above normal we would see a doctor, and at three or four degrees warmer, we'd go to the hospital. And please don't forget, if the world warms 4°C, the planet will already have suffered the impacts from warming 2°C degrees, then 3°C, just a few decades earlier.

Of course, what matters most is not what the thermometer reads, but how nature and humanity fare in a warming world. It's easier to predict biophysics than human beings; what ice or photosynthesis will do in response to a given temperature change can be tested experimentally, while the response of human institutions and societies under completely novel conditions cannot.

Some of the best-studied places are the wealthiest. Please keep in mind that impacts are much worse for the people who can least afford them, like the 2.5 billion smallholder farmers who rely on their farm labor and income to feed themselves and their families.

Fair warning: Climate heating is bad. It ranges from really bad to unimaginably, catastrophically bad.

Every Fraction of a Degree Matters

Humanity is not currently remotely on track to succeed in limiting temperature increase to 1.5°C, the aspirational goal of the Paris Agreement; at current emissions rates, we'll warm past this limit as soon as 2030. Stabilizing temperatures around 1.5°C would require enormous, heroic, sweeping transformations across sectors and societies, plus some luck (more on that shortly). Reaching a 1.5°C world would be an epic victory for humans and nature. It would help keep low-lying cities and even countries on the map ("One point five to stay alive" was a rallying cry from small island nations in Paris). Compared with a 2°C world, it would provide three times as much habitat for insects (the basis of many food chains) and twice as much for plants (the basis of nearly all food chains) and vertebrates (picture your favorite stuffed animal as a kid). And in a 1.5°C world, cleaner air would massively improve public health: clean energy would be widely accessible to help lift people out of poverty without increasing pollution.

Even so, for living creatures, a 1.5°C world is no picnic. Warming is already stressing sensitive ecosystems, and further warming will cause some unavoidable loss and damage compared with today. One of the most unique and threatened systems is coral reefs. Warming ocean temperatures are causing repeated and increasing bleaching worldwide, killing the tiny creatures that give dead coral skeletons their vibrant living skin. Bleached corals first become ghostly white before they're overgrown by a slimy blanket of algae.

My colleague Kim Cobb, a climate scientist at Georgia Tech, describes her research dive in 2016 through the dead coral reefs of Christmas Island in the remote Pacific as her "wake-up call that we are simply out of time." It was on this dive that she realized that the impact from tropical ocean warming "we thought was maybe a couple decades out . . . was actually at our doorstep. In fact it had actually come in and sat on the sofa, and made itself at home. . . . These reefs . . . would take 40 to 50 years to recover . . . [they're] not going to have that kind of time. This is happening much faster than I thought it was going to. I'm devastated at the loss of this very fragile and vulnerable ecosystem. But we're going to be next." Even if we succeed at limiting warming to 1.5°C, only 10 to 30 percent of coral reefs will remain alive.

That's a lot better, though, than the virtual elimination of coral reefs worldwide expected under 2°C, a threshold we expect to pass by 2060 under current climate policies (on the way to 3°C-plus). Nearly half the world's current population, 3.6 billion people, will be exposed to water stress under 2°C of warming. A 2019 study found that with 2°C warming, US residents would feel as if their hometown had moved an average of about three hundred miles (five hundred kilometers) south. For example, Minneapolis would feel like Iowa City; Boston like a suburb of Baltimore; and New York City like DC. Residents of Tampa would experience temperatures similar to those in a suburb of Mexico City. Unsurprisingly, considering how much half an Ice-Age worth of warming would redistribute climates over the globe, my research with colleagues found that more than half of today's wine regions would no longer be suitable to grow their traditional grape varieties under 2°C of warming. Much more disturbingly, basic food production would also be both tenuous and variable, which is a bigger worry, as I hear it is possible to survive without wine; still, I would prefer not to have to try.

Existing policies are expected to lead to warming of about 3.5°C by 2100, after warming to about 3°C by 2070 under no policy. A 2020 study led by Chi Xu found that under *global* warming of 3°C, the average human would experience more than twice as much warming, because people live on land, which warms much faster than oceans. Thus, in a 3°C warmer world, the average person would experience a level of warming similar to them having moved from cool Vancouver to sticky New York City. With 3°C of global warming, nearly 20 percent of land on Earth (mostly in the tropics) would experience temperatures today only found in the hottest 1 percent of the Earth—like the harsh and desolate Sahara Desert, an area that is about the size of the United States but that supports 1 percent as many people. This much warming would make much of the Earth unrecognizable to its current inhabitants. The last time the planet was 3°C warmer was at least 3 million years ago; it was a different enough world that camels roamed north of the Arctic Circle.

If humanity continues to fail to make a rapid transition from dirty to clean energy, or if we make a halfhearted transition and are unlucky with how the planet responds to our climate pollution, we might well experience warming of about 4°C by 2100. (Even more warming is not out of the question.) Our wine study showed that 4°C of warming would mean 85 percent of today's beloved wine-growing regions would be unsuited to produce their signature wines. Much as I love wine, I've gotta say that a wine shortage is going to be among the least of our problems on a +4°C planet, where much of life at sea and on land will be dead or dying. Food will be in chronic fluctuation and shortage; the IPCC projects "sustained food supply disruptions globally." I don't mean just your beloved avocado toast, although avocados will be toast. I mean the basic staple grains that do the job of preventing mass starvation. On this overheated planet, the physical toll of heat stress, where it is too hot for the human body to be able to sweat to cool itself, will make activities of daily living impossible across much of the world, including India, Northern Australia, and the southeastern United States; global labor productivity is estimated to be cut nearly in half. In short: A 4°C world is one of immense suffering and death.

A Game of Jenga We Cannot Afford to Lose

I know it's brutal, and I promise I will soon turn to what we can do about it, but I'm not quite done describing why you should be freaking out about a warmer world. That's because climate change is dangerous in two ways: steadily, and all at once. The gradual increase in temperature drives steady impacts in lockstep, but impacts measured by fractions of a degree are not the whole story. The many delicately intertwined parts of the climate system do not always respond in an orderly, linear way to the sudden wrenching of the thermostat of the whole planet.

We haven't yet talked about the abrupt dangers from "tipping points," where a small increase in temperature crosses a threshold to a dramatic, essentially irreversible system flip, like the dieback of the Amazon rain forest, or the Greenland ice sheet disintegrating. The climate website Carbon Brief likens tipping points to a game of Jenga. The system gets more deformed with each block removed, representing the steady increase in temperature; the last block that makes the tower come crashing down is the tipping point. As in Jenga, the likelihood of a crash increases with each perturbation, but it's impossible to know exactly which block is the last until it's too late.

Some tipping points are self-reinforcing through feedbacks that amplify warming in a vicious circle. The more carbon humans burn, the more we flirt with nature unleashing more carbon of her own. For example, today there is twice as much carbon in soils as in the atmosphere. Much of this soil carbon is stored in permafrost in high latitudes, where it is currently too cold for microbes to break it down. But a bit more warming could bring these frozen soils to life and release millennia of stored carbon to the atmosphere over a few years, driving more warming, which will release even more carbon.

The 2018 IPCC report assigned a "high" risk to tipping points around 1.5°C warming, which the world is likely to see in the next decade or so. This is something I lose sleep over.

Distinguished scientists led by Tim Lenton of the University of Exeter wrote a comment in November 2019 that basically screamed from the pages of *Nature*, the most prestigious journal in science: "Consideration of tipping points helps to define that we are in a climate emergency . . . both the risk and urgency of the situation are acute." They acknowledged that the science of tipping points is not settled but concluded imperfect knowledge was no excuse for inaction: "Given its huge impact and irreversible nature . . . to err on the side of danger is not a responsible option. . . . Warming must be limited to 1.5°C. This requires an emergency response."

In my line of work, I read a great many scary climate studies, but the ones I have the hardest time getting through, that I have to steel myself for the most, are about catastrophic sea level rise, which could be triggered by passing ice-sheet tipping points. It fills my stomach with lead to imagine drowned beachfront homes and whole coastal cities and even low-lying island nations being wiped off the map.

Sea level rise illustrates both the steady and dramatic dangers of climate change. On the slow and steady side, warming expands water already in the oceans (thermal expansion) and melts ice on land into water that eventually flows to the ocean. Sea level has already risen in the twentieth century, driving higher and more frequent coastal flooding. Parts of New Orleans are already being abandoned to rising seas; Miami is not far behind. More sea level rise is already baked into the system; a further twenty- to thirty-centimeter rise by 2050 is broadly agreed. The IPCC concludes that it is *"virtually certain* that global mean sea level rise will continue beyond 2100, with sea level rise due to thermal expansion to continue for centuries to millennia" (it takes a long time for the heat in the slow-moving oceans to catch up with what's added to the faster-mixed atmosphere).

If humanity fails to prevent major sea level rise, coastal cities will be profoundly changed, to the extent that they may have to be abandoned. Authors of a 2019 sea level study conclude that "even in the US, sea-level rise this century may induce large-scale migration away from unprotected coastlines, redistributing population density across the country and putting greater pressure on inland areas."

The National Oceanic and Atmospheric Administration has an online sea level rise map linked to local plans; the scale runs to ten feet of sea level rise, which is too low to capture some of the "extreme" scenario results. Under their lowest scenarios, by 2100, San Francisco is planning for one and a half feet of sea level rise; New York City expects two feet. The changes under this amount of sea level rise are barely visible on NOAA's map.

But you can't miss the difference between today and the second-highest (though short of the highest, "extreme") scenario, where San Francisco expects more than eight-feet-higher water. This would drown city icons including Fisherman's Wharf and Pier 39, the Ferry Building, the baseball stadium, and many of the tech headquarters, coffee shops, and craft breweries around the Caltrain station.

Under their second-highest scenario, New York City expects nine feet of sea level rise, which would bury much of Lower Manhattan, Battery Park, and the East River shore in Manhattan; all the piers; and basically all of Red Hook and the Navy Yard, in Brooklyn.

If we fail to stop climate breakdown in time, I worry that the slow erosion of Earth's life-support systems, overlaid with ever-escalating climate catastrophes, will cause the increasing failure of human institutions stretched beyond their breaking points. After running out of food and water and other basic needs, how long can goodwill and democracy last? There is a limit to the planetary warming that human civilization can tolerate and retain our humanity. It's impossible to predict with precision what that limit is, but it's possible we will exceed it on our current emissions trajectory. I don't want to find out. I don't want to witness the reversal of the long arc of history bending toward justice and watch it bend toward material and ecological impoverishment instead. I don't want to have to look today's young people in the eyes and tell them to their face that, even knowing the stakes and the urgency, we chose, both explicitly and through willful ignorance, to fail them.

What is at stake under the different degrees of warming we face right now is nothing less than the progress of the whole human endeavor. There is much more work to do until humanity succeeds in providing equality and sufficient opportunity for everyone. Still, enormous social and human progress in health and education and equality have been advancing through hard-won struggles. Global warming in my lifetime has the potential not only to undo much of the last century's progress but also to permanently extinguish the very idea of human progress. It is not reasonable to expect or hope that things generally will be better for our children than they were for us if they live on a dying planet.

Welcome to Club Climate Alarmed

Time for a deep breath. In this chapter, I've given you an insider's tour of the terrifying gauntlet that is humanity's potential near-term climate future. You're probably freaking out. Welcome to the club: None of the scientists I've privately asked about their Climate Freakout Level from one to ten rated themselves lower than a seven. This is really hard news to take in, but we have to recognize the severity of the mess we're in before we can start responding to it appropriately.

But so far, I've only told half the story. We've covered the damage that the Exploitation Mindset is doing to the climate, but now we need to lower our sights down to Earth and take a look at how we're treating our land and oceans and the creatures who live there. This too will not be pretty. But please hang with me for just one more super-depressing chapter. Once we are up to speed on the facts, I promise, we will start to run toward responses and solutions as if our lives depend on it, which they do.

Chapter 2

We're the Asteroid

Earth Will Be Fine—But Humans Are in Trouble

Where is the place you remember the feeling of sunshine on your face as a child?

For me, it is the beach at Gualala, a wild spot on California's Mendocino coast, three hours north of San Francisco. My family went to Gualala every summer, just like my dad had done as a kid. After filling my pink knee socks with burrs from a short hike through a weedy field, descending a cliff on rickety wooden steps my grandfather built but no one maintained, and navigating the jumble of tree trunks at the mouth of the little creek that emptied there, we reached a cove of golden sand with craggy rocks offshore standing guard to the Pacific.

My dad would take my sister and me to tidepools full of beckoning mint-green sea anemones. We couldn't resist yelling "Poke!" as we jabbed our fingers into their soft center and felt the suction of their tentacles trying to engulf us. We would delight in catching hermit crabs that had taken up residence in washed-up shells and wait patiently (never my strong suit) for them to stir to life, grip their thin legs into our palms to right themselves, and scuttle off

sideways across the sand. A gleaming abalone shell from Gualala
the size of a baseball cap sits beside me now as I write.

As a teenager, I was called to another landscape shaped by the
Pacific Ocean, but a couple hundred miles inland. I spent a lot of
time hiking and camping and climbing in the Sierra Nevada. This
mountain range was formed by an oceanic tectonic plate pushing
under the continental North American one to create mountains
that rise from the farmland of the Central Valley, through gold
rush foothill towns, to stark, towering granite landscapes like the
spectacular Yosemite Valley. I spent a summer at the end of college as a camp counselor near Lake Tahoe; my friend Pubby and
I joked that summer "ruined our lives" because nothing that followed could live up to its perfection.

I love how expansive the mountains feel. Simplifying life down
to the necessities I can carry on my back forces me to slow down
and notice details that I would normally rush past, like sunlight
catching dew on a spiderweb. I love the feeling of a vista slowly
slipping into view as I climb a mountain. I will never forget watching the sun rise over Lake Tahoe and Fallen Leaf Lake, glinting
off the expanse of snowcapped peaks behind us in the Desolation
Wilderness, after hiking up Mount Tallac by the light of the full
moon with some of my dearest friends.

I care about climate and ecological breakdown because they
harm the people and places I love. Rising seas and stronger storms
are scouring away my childhood beach. There are no more big
abalone shells to be found, because they've been decimated by
overfishing, weakened by pollution and disease, and starved as
their only food source, California's kelp forest, followed other mass
mortalities along the food web in warmer oceans. The hermit
crabs will have a harder time finding a home, because the chemistry of the acidifying sea makes it harder for shell builders to

assemble their armor. Precipitation in the Sierra Nevada is falling more as rain and less as snow in warmer winters. Reduced snow-pack makes my hard-earned mountaintop vistas more monotonous, and it also means California farmers, who grow nearly two-thirds of the fruits and nuts in the United States, have less water to get through the longer, hotter growing season.

I care about stopping climate change because I care about the landscapes and nature I grew up with, and the traditions, culture, and way of life of my hometown and community. These are the tangible ways that climate change affects my lived experience, my identity, and my relationships with the people and places I love. I only know who I am in the context of the climate that shaped and underlaid all my experiences, that made my choices possible. When climate changes the planet, my existence and identity are at stake too.

My specific concerns reveal my personal passions and the privi-leges I've had to enjoy beautiful beaches and mountains, where I feel most alive. But these examples are part of a much bigger story. Whatever it is that you care about, wherever your Gualala is, the people and other species who live there are already being harmed by all the pressures degrading and destroying life on land and at sea and heating our climate. This is a personal tragedy for your favorite holiday spots, but it's also a deep problem for humanity, because ultimately, nature is not a luxury or a nice-to-have. Na-ture is life itself, and the means needed to sustain it. There is no substitute for the fundamental building blocks of life. To meet our most basic human needs, we are utterly reliant on nature. The Exploitation Mindset is putting lives at grave risk of more harm, because without enough healthy nature to support and sustain us, people will suffer and die. (I told you this chapter wouldn't be pretty!)

Today's "Normal" Isn't Normal

A beautiful and tragic thing about being human is that each generation meets the world fresh. For each of us, the world is what we were born into and grow up with. Everything we learn comes either from our direct experiences with and observations of that world or from what someone else (a parent, a teacher, a book) tells us about it. This renewing of collective consciousness can be wonderful, as fresh minds question limiting assumptions or expand past old prejudices.

But this collective innocence can also be devastating, because each generation perceives the degraded condition of nature as the normal condition. It can be hard to comprehend how staggeringly people have hobbled life on Earth, because our baseline for comparison is constantly shifting toward the increasingly impoverished world we live in. We think it's normal that humans are driving species to extinction at a rate *one thousand times* faster than natural; it is not. No one born after 1985 has lived through a normal year on planet Earth; every year of their lives has been warmer than the twentieth-century average. But a study led by Frances Moore at the University of California, Davis, showed that people just stop remarking on warmer weather after about five years. Although extreme temperatures still made people miserable, they stopped talking about it, seeming to accept it as the "new normal" even if there is nothing normal about it. People quickly forget what normal is, what the true baseline should be.

To go beyond our own personal experience and put our human lives in the context of the tapestry of life on Earth, at the geologic scale of the planet itself, we need to rethink our relationship with nature and the physical world. This means we need a different yardstick of time to measure our lives and their impact. Juxtaposing our lifetimes against the lifetime of the planet shows both how tiny and how powerful humans are.

Our Place in the Universe

Let's face it: Our planet's days are numbered. Around a billion years from now, as our sun burns through the hydrogen fuel at its center and increases in size and brightness, it will raise the average temperature on Earth above 100°C, boiling off all our oceans. Eventually all life on Earth will certainly be destroyed, and the sun will probably engulf the planet itself, although it will barely register the incorporation of everything nature or time or humans have ever built as it completes the galactic life cycle of countless stars.

Clichéd as they are, the truisms are, well, true: Change is the only constant. Nothing lasts forever. Out of the mind-boggling vastness of the observable universe, across 93 billion light-years of distance and 13.8 billion years of time, we live at the only address we know of where the lights of consciousness are on. This does make one pause to consider how we humans are currently treating one another and the rest of our fellow life-forms on Earth.

It has taken billions of years for evolution to work its magic through the march of countless generations and give us the stunning variety of creatures bringing the land and seas to life. If we imagine that every living creature writhed or walked or hopped or swam or flew onto a cosmic bathroom scale, their biomass in gigatons of carbon, that currency of life, would give us one very, very crude measure of the richness of life on Earth. Now, we should not get hung up on this one number from a scale; it tells us nothing about the beauty or uniqueness of what's there, but it does give us a first picture of the life we share the planet with, and our place among it.

At this cosmic weigh-in, we need to give a shout-out to green: Earth's dominant life-form, plants, constitute 82 percent of biomass. A distant silver medal goes to simple, unicellular bacteria, trailing behind at about 13 percent of biomass. Bronze medal goes to fungi, from oyster mushrooms to the yeast that gives us bread, wine, and beer (thank you, yeast!); they are 2 percent of global biomass. Archaea, microbes that live everywhere from extreme seafloor vents to wetlands and our guts, where they produce the methane in our farts, are 1.3 percent. Next come protists, including algae and amoebas, at 0.7 percent.

What's left for animals, the least likely kingdom, is a rounding error of life on Earth: just 0.4 percent of total biomass. Most are invertebrates. Nearly half of animal biomass today is in the crunchy exoskeletons and delicately jointed legs of insects and their kin, a quarter in the slippery bodies and glinting scales of fish. Eight percent crawls beneath our feet, as worms slowly bring air and life to soil. Another 8 percent are mollusks, found mostly in the oceans, from twenty-meter-long giant squids to opalescent abalone. Around 4 percent are jellyfish, corals, and their relatives. Most of the rest are mammals. About two percent of animal biomass on Earth is you, me, and the rest of humanity; when viewed on the cosmic weighing scale, our species is about 0.01 percent of all life.

Since this spectacular array of biomass is all known life in the universe, and all of it is aboard Spaceship Earth, as Buckminster Fuller called it, life itself might cease to exist when the sun boils off our oceans. However, our billion-year heads-up gives us a good long while to become a multiplanet species or find some extraterrestrial pen pals to accept our last will and testament.

We're the Asteroid

Until Earth is engulfed by the sun, however, we know at least some life will survive climate change and other conceivable disasters. As it has for about 4 billion years, evolution will work its slow, undirected magic no matter what, tinkering with whatever genetic material is at hand to (re)populate the seas and land with fantastic creatures. It's done this through five (!) previous mass extinctions that wiped out up to 96 percent of life on Earth, triggered by cataclysmic natural forces like massive asteroid impacts or volcanic eruptions.

Today, humans have become cataclysmic natural forces unto ourselves. As Elizabeth Kolbert writes in *The Sixth Extinction*, our snuffing out of many of our fellow species is likely to be humankind's most enduring legacy. We're not the first species to transform the planet with chemistry, but we're the first ones to do it consciously. As the biologist Stephen Porder writes, first blue-green algae 3 billion years ago, then land plants about half a billion years ago, had global impacts through their phenomenally successful oxygen production and carbon drawdown, respectively. While it's hardly fair to hold pond scum and ferns responsible for the ways in which they changed the world, we humans *know* that we are causing profound damage with our climate pollution and destruction of nature. Despite our hubris, however, we do not have the power or capacity to destroy *all* life. Just most of ourselves and the animals we love. Yay!

We've enormously accelerated the rate recently, but humans have been wiping out life on Earth for a *long* time. The killing spree known as the Quaternary megafauna extinction escalated fifty thousand years ago, as humans spread across the globe and hunted large animals and competed with them for space and food.

My colleague Søren Faurby, a vertebrate zoologist, has mapped where mammals used to live around the world before humans killed them off (potentially aided by environmental changes). I was stunned to realize that pre-human Texas would rival the Serengeti; it would naturally be an epicenter of diversity for enormous creatures like the thousand-pound, armadillo-like glyptodon whose Ice Age skeleton is on display at the American Museum of Natural History in New York. Massive elephant-like mammoths ranged across North America, browsing near what today would be New York City, Chicago, Tampa, San Antonio, Seattle, and Los Angeles. Okay, fine, it might be a bit chaotic to have elephant-size critters literally roaming modern city streets—but I'm still pissed our ancient ancestors extinguished any hope for me to ever see these marvelous beasts.

Amazingly, the effects of human-driven extinctions of large wild mammals thousands of years ago, and the fragmenting of the populations that remain, can still be felt today; areas where big animals were wiped out long ago are now hotspots for infectious disease emergence. Chris Doughty from Northern Arizona University hypothesizes that disease-causing pathogens are more likely to become highly infectious and attack new hosts (like people and our cattle) when they evolve independently in smaller, isolated populations of the animal hosts that remain, rather than being spread out among a large, intermixing population. Before our ancestors invented agriculture, their killing spree may have set the wheels in motion for more frequent and devastating human and livestock pandemics.

Modern human exploitation of nature is also increasing pandemic risks. People come into contact with new animal species as they destroy habitat through logging, mining, road building, and other activities. As humans expand into the degraded habitat that more easily transmits disease, the disrupted animals that remain may harbor diseases searching for new hosts—and we fit the bill. Sixty percent of emerging infectious diseases originated in (mostly wild) animals and were transmitted to humans. Most recent pandemics in people were caused by diseases that originated from animals, including COVID-19.

By three thousand years ago, when standard alphabetic writing was spreading around the Mediterranean, half of large land mammal species (creatures about the size of a hefty Labrador retriever and up) had been driven to extinction. This trend of humans wiping out a lot of the big, cool critters in the wild, and thereby favoring the small, crafty ones who can live off of or among us, has been called "the ratification of the Anthropocene." In the Age of Man, the space we make for nature goes to rats.

Going back only a handful of generations shows how quickly and completely humans can erase whole natural wonders. From a lecture by William Cronon, the environmental historian at the University of Wisconsin–Madison, where I did a master's degree, I learned that the passenger pigeon was once the most abundant bird in North America; it was constantly on the move throughout the eastern forests and the Midwest. The naturalist John James Audubon attempted to count the birds passing in an 1813 migration but soon gave up, as their numbers were too enormous. He wrote in his *Ornithological Biography,* "The air was literally filled with Pigeons; the light of noon-day was obscured as by an eclipse. . . . I cannot describe to you the extreme beauty of their aerial evolutions." After a full day of traveling more than fifty miles with no end to the flock in sight, Audubon reported, "The Pigeons were still passing in undiminished numbers and continued to do so for three days in succession."

But only a few decades after intensive hunting began, the very last passenger pigeon, Martha, died in the Cincinnati Zoo in 1914. I remember sitting in that dark lecture theater, peering at the black-and-white photograph of Martha, and feeling a pang of loss. It's strange to miss something I never even knew I lacked.

Looking at life on the planet as a whole, where does this leave us today? What our cosmic scale would tell us is that, at a first approximation, humans have made the Earth half as alive as it used to be, since most biomass on Earth is plants and we've cut total plant biomass in half. Almost 40 percent of forests globally, and more than 70 percent in tropical Asia, have been felled and burned and cleared, mostly to make way for our crops and livestock. At the same time, deep-rooted grasslands that accumulated carbon were dug up for our quick-growing seeds to cycle through the top few inches of soil every few months. Through land conversion, food, and forestry, humans appropriate a quarter of the carbon that would otherwise be accumulated by plants each year. Researchers call this "an integrated socio-ecological indicator of the magnitude of human colonization of global ecosystems."

Focusing down on the animal kingdom, human devastation has been even more profound; people have rapidly depopulated the land, seas, and skies. Only 15 percent of the original biomass of all the wild land mammals remains. The *Living Planet Report* monitors wild vertebrate populations since 1970. In the course of writing this book, I had to update their findings from a 60 percent decline, reported in 2018, to 68 percent in 2020. I had written that 60 percent was devastating; 68 percent is catastrophic. I am running out of adjectives.

A 2018 study showed it will take *millions of years* of evolution for mammals to recover the biodiversity that has already been lost. If we gathered all the wild mammals left on one side of our cosmic scale—all the elephants, tigers, hippos, and zebras; all the buffalo, bears, and wolves; all the deer, squirrels, and mice; all the whales—humans would outweigh them by eight times. If instead we gathered all our livestock, especially cows and pigs, on one mooing, oinking, stinking side of the scale, and put the wild mammals from land and sea on the other, it would be even more uneven: Our barnyard animals now outweigh wild ones nearly fifteen times.

The animals that remain are a pale shadow of the teeming life of even the recent past. Fishermen in Key West, Florida, today are thrilled when they catch a foot-long snapper; they hang their prize-winning fish with pride on the display board. They don't realize that the prevailing champion catch in their grandfathers' day was a six-foot shark that would have literally eaten their fish for breakfast. In an ingenious study, Loren McClenachan used photographs from library newspapers of trophy fish on the same dock over the decades to show how the catch has gotten almost 90 percent smaller since 1956. People still pay the same price to catch tiny snappers today instead of massive predatory fish.

The photos from this Key West dock reflect a larger trend. The comprehensive 2019 global biodiversity assessment found over-fishing was the biggest cause of the decline of nature and biodiversity in the oceans (with habitat destruction and degradation, climate change, then pollution rounding out the pummeling). Globally, one-third of marine fish stocks in 2015 were overfished. Almost all the rest are already maximally sustainably fished, and the Food and Agriculture Organization of the United Nations concludes there is little scope to further expand fisheries.

It's not just at sea that the limits of exploitation are being reached. The rapid increases in technology, wealth, and industrialization as part of "The Great Acceleration" since 1950 have driven enormous human exploitation of the Earth. As a result, the 2019 global biodiversity assessment pronounced nature to be "deteriorating worldwide." Humans are destroying nature to overextract crops, timber, and other materials, stressing and overwhelming nature's ability to sustain that material production, much less clean up after us. Fully 85 percent of wetlands (which filter water and provide flood protection) and half of live coral cover on coral reefs (which buffer storms and are nurseries to much of the life in the oceans) has already been lost. My colleague Ken Caldeira likens our current relationship with nature to a "tragedy" where "because we want to be a few percent richer, we're willing to lose all of these ecosystems. . . . It's like somebody saying, well, I have enough money, so I can run through the Metropolitan Museum and just slash up all the paintings."

On the universe's calendar, life on Earth will go on for a very long time before it's snuffed out by the sun, even if we continue to destroy nature and overheat the climate. But life on Earth as humanity knows it is at risk *right now*. We're the asteroid.

Why We Need Nature

My childhood dream was to be an astronaut. As a kid, I covered my walls with NASA photos of unexplored worlds. Yet astronauts who train for decades to reach the heavens report being most awestruck by looking back at the beauty and fragility of this world we share once they leave it. From space, the only planet with breathable air and drinkable water stands out in sharp relief against the black, lifeless void.

So far, at great expense and significant risk, the best of human engineering has managed to keep just a few hundred people alive outside of Earth, for relatively short periods, a handful at a time. From the first human space flights to today's International Space Station, fewer than one out of 10 million of our planet's current inhabitants has ever left its confines and traveled into space. Humanity is bound up in our shared planetary fate together; leaving, at least right now, is not an option.

Humans haven't even been able to successfully re-create the

workings of healthy ecosystems here on Earth. The Biosphere 2 experiment aimed to support eight participants for two years in a self-contained bubble in the Arizona desert but had to be cut short for safety when oxygen levels plummeted and crop yields declined after their pollinators died.

Human life utterly depends on the variety and health of our living kin and the forests, wetlands, and other ecosystems where they live. In addition to being pretty badass in their own right, these are our life-support systems. We need these healthy and functional ecosystems to keep water, air, and nutrients circulating in order to have food on the table.

We need the humility to acknowledge that humans do not have the power to create the underlying conditions to support our civilization. In our laudable quest for discovery and exploration, we cannot forget that we must first and foremost conserve the only planet that does. Quite simply, we need a stable biosphere and a stable climate to support anything recognizable as human civilization—here or, certainly, in space. The human creations that add up to the whole of "who we are"—from education, democracy, science, and innovation to freedom of the press, social safety nets, culture, and the arts—have largely evolved under the relatively stable climate humans have enjoyed during the last ten thousand years. And these uniquely human constructs can only persist under conditions where safe and stable environmental conditions support and enable them. These are what's at stake as we warm our climate and sicken and destroy healthy ecosystems.

Home Underfoot

People come from places. Our places get into our bones and shape who we are. Like wine, we have the fingerprint of where we were raised etched into us, our own personal terroir. It's what feels like home to us. It's the angle of the sunlight you grew up with. How it looks out your window on Christmas, whether it's snow piled up to the windowsill or a warm sunny day on the beach in Sydney. The first bright yellow daffodils from the garden after a long winter, lazy days by the water in the summer, the smell of the grape harvest in the fall, a warm mug of glögg at the Christmas market.

As an immigrant, when I moved to a new country, I quickly learned that the first question people asked me was inevitably: Where are you from? As soon as I was identified as "not from here" by my accent or appearance or behavior, my fellow humans wanted to know my place of origin in order to make sense of me. I feel more and more at home in Sweden now that I recognize the passing of the seasons through the flowers that bloom, following nature's calendar; now that I know where to look to find the first signs of spring or which trees to detour past in the fall to catch their fiery splendor.

From my friend Mehana, I learned that Native Hawaiians introduce themselves as part of a long line of ancestors from a particular place, naming the peaks and reefs that feed their bodies and spirits. I participated in a similar custom in a scientific workshop in Stockholm where we were asked to draw our "river of life" on a piece of butcher paper, to tell the story of how we came to be here doing this work. My fellow scientists told stories of people who inspired them and places that recharged their souls; together, that was the story of who they are.

Studies show that direct, repeated experiences with local nature over time is how people, especially children, build a relationship with nature and a sense of place and feel a connection and responsibility as well as agency to protect nature. Spending regular time digging in the backyard garden, visiting the local park, or meeting friends for a quarterly beach cleanup have lifelong impacts. These experiences are much greater gifts than a once-in-a-lifetime tropical dive or a photo op riding an elephant through the jungle. It's the same way you build up a lifelong friendship, with many small encounters of increasing intimacy over time, as opposed to partying for one crazy night with strangers you'll never see again.

Humans need a personal, physical connection to natural places to feel our most alive and grounded. People care about particular places, where we smell the salt in the air and feel the sand squidge between our toes, year after year as we deepen our connection to this particular corner of the world. I will always be able to picture the cove in Gualala and feel the wind in my hair and the sun on my face; this connection is something I take with me wherever I go. We need this direct experience with nature, with our fellow life in the place where we live, to understand the web we're a part of and to start to embrace our responsibilities of care in turning from exploitation to regeneration of the living world.

Chapter 3

Uprooting Exploitation, Sowing Regeneration

Time for a New Mindset

Let me sum up the science I've shared with you so far: There are physical, chemical, and biological boundaries to our home, planet Earth. Humanity's current consumption of natural resources (particularly the overconsumption from those of us in the top 1 billion globally) is increasingly exceeding these planetary boundaries, causing a breakdown of our life-support systems. This is a Very Bad Thing.

How did we get to the edge of this cliff? It comes back to the idea of the Exploitation Mindset from the introduction. The Exploitation Mindset sees producing and consuming more material goods as the path to progress and purpose. This is achieved through maximizing speed, efficiency, convenience, and profit, often through technology, which tends to centralize power; the result is an ever-increasing flood converting resources into waste. Following the Exploitation Mindset has been a one-way street turning fossil fuels into vast economic wealth and crushing ecological poverty and debt.

In the Exploitation Mindset, rather than working with nature, man [*sic,* but not really] is at war with nature. It's a war to extract as much material as can be turned into money as fast as possible. On a finite planet, where the rate of extraction is overwhelming the inherent renewal times of natural systems, this is not a war that humans can win. The path we're on—relentlessly increasing production at any cost, racing to be the first to exploit every new resource—ends with cutting down the very last tree and pulling the very last fish from the sea. This would be the opposite of victory.

Treating nature like an enemy to be conquered, tamed, and vanquished is a terrible idea.

With the scope and scale of our consumption, humanity is now a big world on a small planet, as Johan Rockström says. We don't have the option to rewrite the laws of physics and chemistry and biology; that pesky material reality is not subject to election cycles or polling data. Human impact is now so big at the scale of the whole planet that we cannot keep treating the atmosphere as a free sewer for our waste without destabilizing the climate and therefore our societies. We cannot destroy the homes of countless other species and poison those who remain without unraveling the web of life—a web that includes us.

What we *do* have an option to change is the stories we tell ourselves about who we are as individuals and as societies, our place in the world, and ultimately what it means to be human. The Exploitation Mindset holds that nature is a frontier "out there" to be conquered and subdued under human control, manifest destiny written over the whole planet by heroes who tamed the desert. According to this story, we have succeeded; nearly 80 percent of land and 90 percent of the ocean is now dominated by humans, and we're rapidly killing off much of the life that lives there. It's time to put the old story of the Exploitation Mindset to bed and to transform beyond its harmful, outdated values.

A New Mindset

Our success or failure as a species now depends on how we learn to live well within some fundamental limits of the biophysical world, by adapting our societies to provide essential needs, rights, and freedoms within the limits of nature that we've met in Chapters 1 and 2. Accepting the material limits of our home planet is not accepting defeat; it is acknowledging reality. As Jonathan Foley writes, limits can be seen through the lens of opportunity and abundance, not just scarcity. These constraints should spur reflection on what goals are worth pursuing, and creativity and ingenuity in how to live well within our means. To avoid degradation, systems should be managed so they are regenerative on *human* timescales, so that we don't take out more stock than nature can replace or produce more waste than nature can assimilate. We have a lot to learn from nature about how to make our societies solar-powered, resilient, infinitely regenerative systems that support life with no waste (the key to nature's success over the last several billion years).

I believe we need to put our core values at the center of this transformation; to do so, we need to be having many more conversations about what it is we value. The debates and discussions about climate that matter now are moral, not scientific ones, about the kind of sky we want to make, and what that says about what it means to be human.

To make this switch, we need a new mindset. What if we chose to live, write, tell, and pass on a new story? What if we saw the human condition as a story of regeneration rather than exploitation? In the Regeneration Mindset, we'd conceptualize "growth" as a process of renewal and restoration and care for what matters most, making ourselves more resilient against harm. Progress from this point of view means finding a way for everyone, now and in the future, to live a good life, within the limits of the biosphere. This is the sweet spot that Kate Raworth calls the "safe and just space for humanity."

The Regeneration Mindset is named for one of life's most incredible properties, its capacity for healing and renewal to carry on, both within individuals and across generations. To regenerate is to be formed again, to generate or produce anew; to "change radically and for the better," to be "restored to a better, higher, or more worthy state." In the Regeneration Mindset, purpose comes from pursuing quality of experience, not just quantity of material; deepening rather than just speeding up our lives.

A core insight of the Regeneration Mindset is to recognize that each of us exists in a web of relationships, with our fellow humans and the living world, to whom we have duties and ties of care. We humans are a part of nature in the same way we are a part of a family; we have our individual identity, but much of it is shaped by our kinship with the larger whole. Life exists not in isolation, but in a web of interactions with its living companions (pollinators, prey, predator; family, friends, communities) and its nonliving environment. Together these living and nonliving surroundings form an ecosystem within which life has a role, embedded in space and time.

Putting the Regeneration Mindset into practice means placing value, respect, rights, and empathy for all living beings at the center of our priorities; recognizing that life is an end unto itself and not a means to an end; and deriving meaning from reciprocal relationships of love and care rather than from transactional exchanges. Regeneration means seeing the Earth as not just the wellspring of resources, but a living entity with whom we have a relationship.

We have to stop seeing nature as subservient to man and to

recognize that nature does not exist to serve us. Evolution is not a ladder or a process striving toward its peak in *Homo sapiens*. With humility, we can see that humans are one wonderful, fascinating, important, and very tiny and recent branch of the evolutionary tree to which all life belongs. People must forge a relationship we can sustain with the fabric of life, which is what nature is. Even if humans didn't need nature for the survival of our species and of the values we hold most dear, it's morally wrong to destroy the complex fabric of life on Earth. The beauty and variety of life deserve to exist and must be centered alongside people's needs in rewriting the story of our civilization going forward.

Three Principles of Regeneration

The Regeneration Mindset will help create the thriving planet and more just world I want to live in and pass on to future generations: where the fundamentally equal value and worth of each person is respected and every person has the opportunity for a life of dignity, regardless of where they happen to be born. A world where everyone's material needs for food and water and shelter, as well as their social needs for connection and care, are met; their rights and voice are respected; and they have the capacity to help shape the world around them through education, employment, and equality. For this world to be born, and for it to be carried forward many generations, it needs to work within the safe operating space of the physical and biological capacity of the planet. I propose three core principles that can help guide us to this world, which I'll return to and elaborate throughout this book:

1. **Respect and care for people and nature.** First and foremost, both people and nature must be respected, because

people are embedded within the biosphere; we are a part of nature. The well-being of both people and nature matter and ultimately depend on each other. People are not just the sum of our brilliant minds and empathetic hearts, but also our wonderful, fragile, organic bodies, made up of material borrowed from stars and soil and soon to be returned to them; meanwhile, we're wholly dependent on nature to meet our natural bodily needs for sustenance and shelter.

2. **Reduce harm at its source, not by treating its symptoms.** True solutions get to the root cause of the problem. To reduce harm to people and nature, trace it back to its origins and look for systemic solutions that stop the harm as completely and quickly as possible, in ways that protect the well-being of people and nature. Harness existing wisdom and creativity. The solution to pollution is not dilution but prevention: designing systems from the start that work with rather than against physics, chemistry, and biology.

3. **Turn our impulse to build toward the building of resilience.** Cultivate a range of options that ensure people and nature can continue to thrive. Resilience, the capacity to recover from setbacks and maintain essential functions, must be at the heart of this new mindset. Resilience in ecosystems comes from health, diversity, redundancy, and connectivity. Resilient systems are flexible; they are adapted to work with rather than against local conditions and to handle the unexpected. They are built from the ground up, starting at local scales, with the strongest relationships with their nearest neighbors, but they have connections and communication that span the globe. These smaller, more flexible systems with more local control are often better adapted to local needs and changing conditions

than massive, centralized, one-size-fits-most approaches. Working within the laws of nature will make human systems more resilient and effective. Conserving all the parts of a system helps preserve stability, function, and options.

Regenerating Nature and Climate

How can we engage the Regeneration Mindset to repair our ecological crisis? We'll need to reduce and rethink overconsumption, and treat nature in a way that maintains and strengthens its ability to regenerate. Protecting remaining intact habitats and forests is a critical top priority for biodiversity and for ecosystem health and resilience, including their ability to take up carbon; restoring degraded habitats is also important.

But we need to be cautious about the Exploitation tendency to see nature as a means to an end. Trees are not machines that automatically store carbon; to do so, they need to be part of healthy ecosystems with their key relationships intact. Research warns that we can't count on sick, stressed ecosystems to solve our climate problem single-handedly, and that planting trees (even a *lot* of trees) cannot undo the carbon pollution of centuries of burning fossil fuels. Planting trees needs to be done in a way that makes sense for local conditions and traditions, to benefit both biodiversity and people and be resilient to future climate change. Finally, tree planting cannot be a smoke screen to keep putting off eliminating carbon pollution at its source. For this reason, Lancaster University researchers caution that we need separate strategies for reducing and *preventing* emissions (essential) and for attempting to use trees (and other "negative emission techniques") to *remove* atmospheric carbon (which may be a necessary additional strategy but is not an alternative one).

How are we going to apply the Regeneration Mindset to stabilize the climate? Fundamentally, we need to get to the root of the problem: stop burning fossil fuels. We need to slash emissions at their source in a way that protects and strengthens the resilience of both nature and people. We need to stop doing things that damage the climate and replace them with things that don't. Overconsumers need to reduce our demand for energy, materials, and food, and use what we do need much more efficiently. Sectors that are not ready to do their share of eliminating emissions by using less fossil fuels (decarbonizing) in the next decade, or where it's technically impossible, like long-distance flights, need to be scaled back and fairly distributed.

Making Fossils History

Right about now you're probably ready for some good news. I've got some for you: What people alive today do *really matters* for the climate. This should light a fire under us to decarbonize as fast as humanly possible, because we are now determining the climate from around 2035 (when we could first start to see signs of our hard work paying off in slowly stabilizing temperatures) and for the rest of human time after that. We know the risks increase enormously for every bit of heating, and we're dangerously close to deadly tipping points. Consider the projected impacts of the world we're heading for right now—drowned cities and countries, coral reefs fading from living memory—a postcard from the future we don't want, one we still have time to avert in the next few years. It's easy to treat possible scenarios for what humans do in the future with a false sense of certainty or inevitability. But what is seen as possible (or inevitable) is changing under our feet. Each coal-fired power plant closed, or never built, makes carbon-free energy a little more possible, closer to imagine, closer to being real.

It is humans causing climate change. It has to be us who fix it. Fixing climate change is simple, but not easy. We know it is primarily our burning of coal, oil, and gas and secondarily our land use and agriculture that are driving climate change. We cannot expect to stop climate change without stopping the source of the problem. To stay within our carbon budget, we simply cannot burn carbon for our energy sources anymore. The Stone Age didn't end because we ran out of stones; the fossil age must not end because we run out of dirty energy, but rather because we decarbonize and make a very fast transition to clean energy.

If we want a stable climate future, we have to leave carbon-based fuels in the ground and replace them with clean ones. Any delays in doing so are attempts to put off the inevitable; all credible roads to a stable climate leave no future for fossil fuels. A study published in *Nature* showed that the quantity of carbon in known reserves of coal, oil, and gas is three times higher than the remaining carbon budget to stay within even the maximum Paris temperature goal of 2°C. Any further development of dirty fuels in the Arctic, or the dirtiest "unconventional" oil and gas (think Canadian tar sands and shale gas from fracking), blows the carbon budget for 2°C. The coal has to stay in the hole and the oil in the soil.

If the climate can't afford burning the carbon in existing reserves, it certainly doesn't make sense to seek new reserves or build new infrastructure that locks in more warming. Yet mind-bogglingly, governments plan to produce more than twice as much fossil fuels in 2030 than fit within a 1.5°C carbon budget, and fossil fuel companies are still spending billions accumulating more unburnable carbon. To meet 1.5°C, a 2019 study led by Dan Tong at the University of California, Irvine showed not only must the world *immediately* stop building anything that runs on coal, oil, or gas (power plants, pipelines, fossil-powered cars), we also need to start shutting down existing fossil infrastructure early. Their business plans simply blow our carbon budget. Sorry, pipelines.

What We Have to Do by 2030

The actions needed to avoid catastrophic climate change are historically unprecedented at the required scope and scale. A lot has to go right in a very short time, and it would have been a hell of a lot easier to get serious sooner, but: It's still possible. Basically, we need to go from rapidly increasing carbon emissions to reducing emissions to zero in my lifetime. We need to cut carbon pollution by at least half, preferably more, in the next decade.

Remember, to stabilize the climate *at any temperature,* we have to *completely stop* adding carbon to the atmosphere. Real zero. If we can cut global carbon emissions about in half by 2030, and zero them around 2040, we have a two-in-three chance to stabilize warming around 1.5°C. If we cut global CO_2 emissions about 25 percent by 2030, and zero them by 2070, we'll have a 66 percent chance of stabilizing the climate around 2°C, the upper limit of the Paris Agreement temperature goal.

To get to 1.5°C, the IPCC states we need deep, rapid emissions reductions in every sector: energy, land, urban, transport, buildings, industry. Given the many sources of greenhouse gases across the world and embedded in every sector, no one technology or strategy will be sufficient alone to approach zero emissions globally with the breakneck speed that's needed. At this eleventh hour, there are no silver bullets left; we will need leaders from every sector, around the world, to step forward. The urgency of emissions reductions means we can't wait around for new tech to save us; we need to start creatively using the existing tech at hand.

Fortunately, we already have the tech we need. A 2018 study in *Science* concluded that most of today's energy services were "relatively straightforward to decarbonize." About a quarter of fossil emissions come from technically challenging sectors like aviation, iron, steel, and cement production, where research and development is needed to bring the costs down, but "combinations of known technologies could eliminate emissions related to all essential energy services and products." While researchers tackle these last-mile problems, we have more than enough to start with today. In fact, there are already lots of pathways toward zero emissions mapped out for industries and countries, and no shortage of creative and effective solutions drawn from sophisticated models and traditional knowledge, from cities and schoolchildren, just waiting to be put into practice or scaled up.

But the transformation needed is massive, and it's not going to be easy. Fossil fuels built the world we live in today. Currently more than 80 percent of global energy heating and lighting our homes and powering our vehicles comes from fossil fuels. Cutting global carbon emissions in half is equivalent to completely eliminating the carbon emissions from today's three biggest climate polluters: China, the United States, and the European Union. Put another way, to cut fossil CO_2 emissions 45 percent, we have to completely decarbonize electricity and heat worldwide. We're talking about tearing out and rebuilding within a few short years an energy network that took centuries to build. (This is a *huge* acceleration of the current rate of building clean energy, which would take 363 *years* to construct the infrastructure needed to limit warming to 2°C, at which point it would obviously be far too late to matter.) We have to go a whole lot faster. The job ahead is . . . not trivial.

But we've simply got to do it, because the alternative is to watch our last chance for meeting 1.5°C, and soon 2°C, pass us by, and to write a story of harm and suffering that life on Earth will feel for the rest of human history. Because carbon lasts for so many generations in the atmosphere, humanity essentially only gets one shot at solving this problem reasonably well, and that shot is now or never.

And on *that* cheery note, congrats! We've reached the end of Part 1, where I've taken you on a whirlwind tour of where humans fit in with life on Earth to show why humanity's future is so intimately entangled with the climate we create. I've argued that it's high time to center regeneration in how we treat the planet and one another, and that decarbonizing as fast as humanly possible and rebuilding our relationship with nature are key tasks for humanity to thrive into the future.

So far, we've faced the scientific facts about the deep scars that the Exploitation Mindset is leaving on the climate and the living world. In Part 2, I'll talk about how I've come to terms with this heartbreaking knowledge personally, from facing my own feelings, learning from and building relationships with my community, and finding inspiration from around the world. I'll make the case for why I believe we will need to harness all these mushy feelings if we're going to truly rise to the occasion of making the necessary, urgent changes toward regeneration.

Part II

We're Sure. It's Bad.

How We'll Get Through Now

Chapter 4

Sink into Your Grief

How to Honor Everything We're Losing

At the southern end of Sonoma Valley, where I grew up, the Mayacamas Mountains end in rolling foothills that sweep down to the north end of the San Francisco Bay. The sheep that used to graze in the marshy wetlands there gave the region its name, Carneros (Spanish for "rams"), but the icon of the region now is wine grapes, especially Pinot Noir.

I chose to study the wine industry that is the lifeblood of my hometown, and more specifically this grape, for my PhD. Vineyards and wine production are deeply entwined with Sonoma's history and have shaped its landscape, culture, economy, and way of life. The unique elements of the soil and climate in a particular place, combined with the region's own growing and producing traditions, give wine and other geographically distinct products like coffee a signature terroir, or taste of place. I studied how growers responded to climate in the vineyard and how chemical compounds that give wine its color, taste, and texture were affected by changes in temperature. These changes alter how the vine regulates its production of sugar through photosynthesis and how it modifies and recombines this one starting ingredient to produce hundreds of compounds, some of which might ultimately smell like cherries or black pepper in your glass.

Carneros Pinot Noir has a signature flavor thanks to where and how it was grown, a combination of bright red fruit and baking spices like cinnamon and cardamom. One grower I interviewed for my PhD research told me, "I opened up the bottle of wine, and it smelled like Carneros. Just that particular kind of salt feel, softness to the air that comes from the Bay influence, and it had some of that smell of the grass and the wind and the sunshine."

Writing those words makes my heart heavy, because I'm afraid I have tasted my last sip of the Carneros Pinot Noir I treasure. The grapes of the cool-climate variety I chose in 2005 to study as an icon vulnerable to climate change are ripening faster through hotter summers, retaining less of their ethereal bright fresh strawberry and tending toward new flavors of jammy plum. I'm afraid I'll never find the unique and irreplaceable taste of where I grew up again, as it gets flattened by climate change.

Something similar recently happened to me in my adopted homeland. In August 2019, my husband, Simon, and I went hiking in northern Sweden. We hoped to climb Kebnekaise, Sweden's highest mountain, though we'd been warned of the possibility of epic mosquito swarms and snowstorms even in late July. We hiked and boated the nineteen kilometers from the trailhead to pitch our tent near the mountain station, waking early the next day to clear, calm conditions: the perfect weather to attempt a summit. We spent that long day in the far north crossing icy rivers that cut through green plains of tundra. As we gained elevation, geology took over from biology, yielding a landscape dominated by dramatic rock and ice.

When we reached the summit, I was elated. I had long dreamed of earning this view, but I'd been having problems with my foot and knee all spring, and I hadn't been sure if my dutiful physical therapy would be enough to let me make the top. I was thrilled, but I was also tired and hangry, and thinking about how late it would be when we got back. The last few meters of the climb to the very tippy-top were covered in icy snow above a sheer cliff; I decided not to risk it. We hadn't brought crampons, which I later read were advised "to prevent you sliding off to certain death" if you fell. I could always go back next year with crampons, I told myself.

What haunts me is that I'll never get to stand on the very top of Sweden's highest mountain, even if next year I'm braver and fitter and better prepared: It's gone. One month after our trip, a geography professor made the same journey and measured the official height of the mountain, confirming that the glacier at its peak had melted enough that the south peak we'd climbed was now lower than its rocky northern neighbor. The maps have to be redrawn; the geography lessons have to be updated. An icon of Swedish identity has changed.

I was drawn to environmental science because I loved hiking in the mountains. In college I discovered I could get paid to do exactly that. My first job in science was as a research assistant to Professor Gary Ernst. Our office was the dramatic, stark moonscape of the White-Inyo Mountains, looking west across the Owens Valley to the snowcapped Sierra Nevada that captured the rain blowing off the Pacific. After mapping the geology of these mountains for four decades, Gary started getting interested in changes he was noticing in the "green stuff covering the rocks." Vegetation communities were marching slowly uphill. Desert species were heading to new alpine addresses at the twisted feet of the oldest living creatures on Earth, the bristlecone pine trees, some of which had been eking out a living from the chalky white soil for nearly five thousand years.

Eventually, I realized that it was impossible to go out in the wild and study nature as a separate entity from people. There is no wilderness left on this planet. Humans have touched every last corner of the Earth, even if no foot has ever been set there, because we're changing the climate, and therefore everything that life depends on. Acknowledging what we're losing to the climate and ecological crises is not easy, but it's necessary to bury the Exploitation Mindset that caused these losses.

Stewards of Grief

Bearing witness to the demise or death of what we love has started to look an awful lot like the job description for an environmental scientist these days. Over dinner, my colleague Ola Olsson matter-of-factly summed up his career: "Half the wildlife in Africa has died on my watch." He studied biodiversity because he loved animals and wanted to understand and protect them. Instead his career has turned into a decades-long funeral.

As a scientist, I was trained to be calm, rational, and objective, to focus on the facts, supporting my claims with evidence and showing my reasoning to colleagues to tear apart in peer review. I was trained to use my brain but not my heart; to report methods and statistics and findings but not how I felt about them.

In graduate school, I was surrounded by brilliant, Serious Men who spoke in even, measured tones about the loss of California snowpack and crop yields; I tried to do the same. I felt my credibility as a scientist was on the line, as was the respect of the men who would sit on my future hiring committee and determine whether I would get a tenure-track job. I internalized the idea that scientists should be "policy relevant and yet policy-neutral, never policy-prescriptive." I was not supposed to have a preference, much less an emotional attachment, to one outcome or another, even on matters of life and death; that was for "policymakers" to decide. (This reticence goes against the wishes of 60 percent of Americans that scientists take an active role in policy debates about scientific issues.)

My dispassionate training has not prepared me for the increasingly frequent emotional crises of climate change. What do I tell the student who chokes up in my office when she reads that 90 percent of the seagrasses she's trying to design policies to protect are slated to be killed by warming before she retires? In such cases, facts are cold comfort. The skill I've had to cultivate on my own is to find the appropriate bedside manner as a doctor to a feverish planet; to try to go beyond probabilities and scenarios, to acknowledge what is important and grieve what is being lost.

Throughout my training as a scientist and my doomed first marriage to my high school boyfriend, I valued my intellect and logic over my feelings, believing that wisdom came from the mind. I thought I could hide from or at least talk myself out of my gut instincts, which often kicked and screamed behind the driver's seat of my rational mind.

Only in the most recent decade of my life have I realized that feelings, manifested as physical sensations in the body like my stomach clenching or my heart lifting, have their own wisdom. I don't have to react to these feelings in any dramatic way if I don't want to; all I have to do is make eye contact, wave, and not run away. Like all feelings, sadness is valid; it need not dictate my actions single-handedly, but it deserves acknowledgment.

I know that there is much greater suffering than my own, like in the low-lying communities in Bangladesh where rising seas are salting their drinking water and threatening their homes, and for the 3 billion animals killed or harmed in Australia's terrible 2019 bushfires. I know that my privilege has shielded me from many hardships and inequities. But I've decided it's pointless to try to compete in the Grief Olympics, that exercise in which I feel I'm not entitled to be upset if someone somewhere else has something much more upsetting going on. It does not diminish the monumental losses to also grieve my personal, smaller ones.

I've realized that giving space to my feelings gives me more empathy with what others are going through as part of the shared human experience and helps me connect with them more deeply. Katharine Wilkinson of Project Drawdown makes a beautiful distinction in her 2018 TED Talk between two responses to loss: a heart that simply breaks, that curls up on the couch and hides away, and a broken-open heart that reconnects with the world around us in empathy and love, a heart that is "awake and alive and calls for action." No matter the object, grief and sadness focus our attention on what matters in our lives, and they turn us into human distress signals: They summon help.

It has taken me a long time to come to terms with my climate and ecological grief, but I've realized that swimming through it is the only way forward. One role environmental scientists can play is to be "stewards of grief, to hold the hand of society as we enter the unknown space of the climate crisis," as my friend Leehi Yona so beautifully wrote on Twitter when the 1.5°C report launched. As scientists, we have had much more time observing the decline of what we love, buried in grief, struggling and sometimes finding ways to digest and cope with it. We are further down the line of where we all must get to as a society, facing hard truths and still finding ways to be kind and resilient, to do better going forward, to get through this together. While we grieve and honor what is lost, we still have so much we love at stake that is worth fighting for.

Swimming Through an Ocean of Grief

I learned a lot about grief when one of my best friends, Pubby, was diagnosed with stage four cancer at age thirty-six. I got the news from his wife and my BFF, Lucy. My mind was reeling as soon as she said, "It's about Paul's health." I struggled to come to terms with medical terminology familiar to these two doctors but foreign to me.

I wanted some sort of prediction, statistics, a number that would tell me something about what was ahead and therefore tell me what I should do now. I called our other BFF, another medical doctor, who knew Pubby's diagnosis already. She was gentle and kind and very, very sad.

I think in numbers for a living. I asked something like, "How long has he got to live?"

I don't remember what numbers Dr. Meg used to answer my question, but I will never forget her human answer: "This diagnosis means we'll be young at his funeral."

She was right, twenty-two months later.

My friendship with Pubby had been sealed over a spontaneous Beastie Boys dance party outside our adjacent dorm rooms in college, where everyone knew him by his childhood nickname, and honed over two decades of philosophical bullshitting into the wee hours. As the lone woman on the groom's side of the wedding party, I hung with the boyz the night before Pubby and Lucy's ceremony. I hoped they didn't notice I kept sneaking into the bathroom to secretly dilute my whisky (sadly, I only acquired a taste for it after he died). Throughout our twenties, Pubby and I had long conversations about how to be a scientist as we each grew into that role. He was clear, though, that facts and science had nothing useful to say about the human condition. For the existential questions, you needed literature and stories.

After that phone call from Lucy, and after Meg so clearly said out loud the heartbreaking truth that helped me understand what was at stake, I changed my plans immediately to come home and spend that summer with Pubby and Lucy and our friends. Impending loss sure does clarify one's priorities. We tried to enjoy simple moments together, not knowing how much time we had left.

We took our last trip to Lake Tahoe together. The familiar Sierra landscape was spectacular, but looking back I realize we were boating on a warming lake that was increasingly filled with algae, past hillsides covered with mighty trees that were dead or dying, desiccated by the ongoing drought that would kill 150 million trees in California and weaken their defenses against bark beetle attacks. We visited our beloved Stanford Sierra summer camp, where a new generation of college students brought the magic to life. Watching counselors raised on *Glee* perform tuneful, earnest pop songs for the talent show, instead of the deadpan absurdist existentialist skits he'd trademarked, he commented, "Huh, I guess staff these days think the talent show is about talent."

Witnessing the end of Pubby's life was incredibly painful, but also full of grace. The head-on way that Pubby and Lucy faced the fact of his impending death made the grace possible. Their experience as doctors made them realize "your own suffering is not special," as Lucy put it. This meant they didn't waste a second agonizing over the question to which there is no answer, "Why me?" (Pubby simply said, "Why not me?" and got on with it.) Starting from acceptance of the reality of the situation, they did their best to get through each day together, constantly reevaluating priorities to maximize meaning under the declining range of what was still possible. As Lucy taught me, "Pain times lack of acceptance equals suffering."

Their sense of shared purpose spilled over to the community of people who loved them, who came by with casseroles and wrote funny notes and watched mindless TV together as Pubby was dying.

His directive that Lucy take care of him, and everyone else take care of Lucy, taught me a lot about how to be part of a community ("lift in, dump out" to give support to your friends closest to the crisis, and vent to those farther away). I learned that "How can I help?" is a terrible text to send to a friend in crisis; "I'm at Target, what do you need?" or just showing up with food or booze or both is much better. It was beautiful to see the incredible way that Lucy took care of him, as a doctor but also a human: anticipating his needs, asking specific questions, giving him choice and dignity over all she could.

Not long after his diagnosis, Lucy booked a catamaran cruise on the San Francisco Bay. It was a gorgeous, sunny, windswept view from the bow, but it was hard not to think about all the lasts and losses and nevers. A man in his sixties stumbled slightly when the boat pitched as he passed us. Embarrassed to appear infirm in front of what he took to be a group of carefree thirtysomethings, he mumbled, "It sucks to get old." We all stared at him wordlessly. No one could muster a witty or polite or kind response to make him feel better, as we normally would, because we were all thinking: "You lucky bastard."

Pubby made confronting his own death unflinchingly his final purpose, and like every other incredibly difficult goal he set his relentless mind to, he knocked it out of the park. As he was writing what would become the book *When Breath Becomes Air,* I confessed I was also thinking of writing a book. His immediate response was, "That's a terrible idea, Kimmy. No one reads books."

Shortly thereafter, two months before he died, he sent me a writing assignment: "2 pages or less on what you are trying to accomplish with your life." I wrote a draft right away. It was terrible: vague, jumbled, more about what I didn't want than what I did. I closed the file and put off revising it. Two weeks later: "You are overdue." I writhed and procrastinated and didn't reply. Identifying and justifying my purpose to my brilliant, beloved, dying friend was hard. Two weeks later: "Still waiting . . ."

I replied, "Ack! Haven't forgotten. Have a draft. Feel pressure to make it good."

"Don't worry about making it good. Make it done." Good first draft advice generally, especially if your editor is actually dying.

I finally sent what I had. His reply was wise, literary, and snarky, and his comments on my muddled draft were incisive and spot-on, as always. My favorite remains, "Anyway, I remain unconvinced as to your larger life goals." I miss that dude.

Facing Inevitable Loss

What I've learned about grief is that it signifies what really matters. An activist slogan inspired by Joe Hill says, "Don't mourn, organize!" I think we need both grieving and action. It is the depth of our feelings that both induces grief and motivates and sustains action. And it is overwhelmingly heartbreaking to realize that we cannot save everything we love.

The first time I heard someone speak openly about climate loss in the here and now was at a climate conference in 2016. I was used to scientists carefully framing hypothetical future scenarios as if a 2°C and a 4°C world were equally attractive buffet options, even if it felt like their results were being delivered through increasingly gritted teeth. So it was a powerful breath of fresh air to see May Boeve, the then thirty-two-year-old executive director of 350.org, say to a room full of older men in suits championing techno-solutions: "We have to be honest, a lot has already been lost, and our hope must be grounded in reality."

It was a relief to be given this permission to acknowledge my climate grief, to admit that there have already been losses and there will be more. Some losses will tear through communities and continents with the speed of a racing fire, and some we have already set in motion but will unfold over millennia. We need to confront the fact that some of our world will not survive, and grieve its loss.

The day I heard about the death of the last male northern white rhinoceros, I was at a museum exhibit featuring the Swedish art duo Bigert & Bergström. One of their works is an enormous sculpture of a rhinoceros. Its body is massive and lumbering and wrinkled, made from iron streaked by rust. In shiny contrast, its silver-colored horn is cast from a northern white rhino horn preserved at the Swedish Museum of Natural History. I realized that museum specimens and artworks were all that was left of this creature, who would never walk this earth again. I felt a wave of grief that I had delayed by swiping past the bad news earlier on my phone. Having an art form to relate to gave me a place to put my grief. It was painful but necessary.

Coastal communities are facing the reality of sea level rise now; climate change is turning their residents into refugees. The capital of Indonesia, Jakarta, is built on drained wetlands, with 40 percent of the city below sea level. Given its vulnerability to rising seas, the government of Indonesia announced in August 2019 it would relocate its capital city about a thousand kilometers away to the island of Borneo.

Louisiana has a new plan to stop fighting the sea and instead move people away from high-risk areas to neighboring towns, aiming to "prepare those communities for the influx already underway." Parishes like Orleans and St. Bernard have already lost more than half their populations since 2000. The price of home insurance in some areas has already doubled or more since 2004.

"The joke is, 'Oh great, now my property will be beachfront property.' We've missed the point that you won't be able to get there, because the roads will be flooded and the power will be out," said ocean scientist Ayana Elizabeth Johnson on the *Mothers of Invention* podcast. "This assumption that we can just build a wall and somehow hold back the entire ocean . . . that's just hubris. It's absurd. We actually do need to move away from the coast for our safety."

Many of the species that are threatened today are the ones we read about as children, that we read and watch films about with our own children. With 91 percent of the Great Barrier Reef bleached in 2016 and extreme heatwaves becoming more common, will we soon live in a world where *Finding Nemo* is as much of a fantasy as *The Little Mermaid*?

My parents took me and my sister to the Great Barrier Reef when I was thirteen, awkward in my first bikini but mesmerized by the vivid, pulsating undersea rain forest. Now coral reef scientists describe diving through tears when they see spectacular reefs they've studied for years reduced to monotonous red-brown algae. Reality is eerily following a progression of photographs from a 2007 *Science* paper: thriving, colorful corals and fish in clear water remaining at 1°C warming; dead white coral plus some algae at 2°C; a dull, uniform slime over dead, broken bits of coral and shell in cloudy waters at 3°C.

Australians who have the Great Barrier Reef as their backyard know viscerally what's at stake. I met Alice, who is originally from Melbourne, on an early summer day at a mutual friend's sunny beach picnic. In the long Scandinavian twilight, we went to the Tivoli amusement park in Copenhagen, the charming and somehow uncommercial inspiration for Disneyland. Since 1843, it's been an oasis in the heart of the city, full of flowers, music, beer, cheerful topless wooden mermaids on the Hans Christian Andersen ride, and general merrymaking of every sort. As we ate tacos in the shadow of the scariest ride, watching the sparkling lights over the lake, she turned to me and said, "The Great Barrier Reef is dying."

I froze. I didn't know what to say. She's right; half the shallow-water corals there have died in recent ocean heatwaves. I wanted to reassure her that everything would be okay. Or that there was hope. Or that the fate of their national treasure could be the motivation for citizens to push their governments hard enough to do the right thing and choose corals over coal. I wanted to offer some sort of scientifically grounded recommendation of What to Do. But I didn't have the right words. I didn't know what to say, so I didn't say anything.

Looking back, what I wish I'd done is not try to give Alice answers but to ask her questions. My decades of training to use my scientific knowledge and analytical brain to diagnose the problem were not helpful in that moment. I wish instead that I'd empathized with Alice, with what it feels like to stand on the edge of losing something important and precious, and that I'd treated that experience as a fellow human. I wish I'd asked her what her earliest memory of the Great Barrier Reef was, what stories she grew up with that made the reef part of what it meant for her to be an Australian, and what she hoped and feared for the future.

Honor It All. Save What You Can.

After decades of avoiding it, I have found it cathartic to grieve climate and ecological losses. It feels right to acknowledge my sadness in mourning a future I expected that will not be. It's like the process for receiving a difficult health diagnosis; readjusting expectations, trying to figure out what is still possible in the new context, what still matters.

But it's not only the losses of life or catastrophic suffering that deserve to be honored and acknowledged through expressing grief. It is also a loss when smaller little happy things like local traditions, cherished heirlooms, and ways of life are ruined. You are not being childish when you feel sad about you and your loved ones losing these elements of your identity. You are being child*like*: seeing the world as it really is, rather than through fogged-up glasses of cynicism. In your sadness lies the power to honor what cannot be saved and to save what is yet salvageable.

The work of grief is the work of facing the loss of what we love. We can make rituals to honor the memory of what was and mourn its passing. We can tell the stories of what is no more to those who come after us, born into a more hollowed-out world. Some of the choices we have today will never be available to the next generation; our choices only existed before they had a voice.

Inhabiting grief means not turning away at the difficult times, but bringing comfort through being in the radical present for the moments we share that may be the last of their kind. Grieving strengthens bonds with our community over our shared care. This is where the Regeneration Mindset can help us ask ourselves, How can I draw on my strengths and the strengths of my community to protect what I care about most, to reduce harm and suffering wherever possible, and to increase resilience so that we survive the tough times ahead with what matters most intact? Doctors cannot save every patient. But each one deserves care, and some can be saved.

When Simon and I returned to the mountain station after climbing Kebnekaise, we saw an art project hanging in the hallway: a banner saying THANK YOU, SOUTHERN PEAK in both Swedish and the local Sami language, signed by hundreds who had skied, climbed, and hiked its peak. It felt painful but important to get to say goodbye.

In August 2019, Iceland residents, including the prime minister, held the first funeral for a glacier named Okjökull. They hiked to rock newly exposed by the melting glacier and installed a bronze plaque. "A letter to the future. Ok is the first Icelandic glacier to lose its status as a glacier. In the next 200 years all our glaciers are expected to follow the same path. This monument is to acknowledge that we know what is happening and what needs to be done. Only you know if we did it. August 2019. 415ppm CO_2."

Chapter 5

Making Meaning in
a Warming World

Personal Choices Under Global Change

Recently I found myself on a video call, giving existential and extremely intimate life advice to an earnest man in the United Kingdom who had e-mailed me out of the blue. Jake wrote:

A couple of months ago I really knew nothing about climate change. I knew it existed, but had no real concept of what it meant in terms of impact on our lives. I have one daughter who is two years old, and myself and my wife have always wanted a larger family. My wife brought up the idea of having another child and I felt a bit of a need to do some research about that nagging thing in the back of my mind that was climate change. I fell down a black hole. Everything I read felt like doomsday was approaching. Everywhere I looked I saw hopelessness. Not only did I begin to question how I could balance my conscience against bringing a new child into this world, only to have to face the collapse of it, but I began to feel a deep regret for my daughter, and what she is already going to have to deal with. Did I make a mistake bringing her into this world only for her to face the worst of it?

Jake and I talked for an hour, about the possible range of climate outcomes during his daughter's life, and what he was already doing and could do to face the climate crisis as a parent and citizen, at home and in the world. He clarified that he wasn't expecting me to make his family-planning decisions for him, but he wanted to hear my thoughts.

I thought of Pubby and Lucy, deciding to have their daughter Cady after his terminal diagnosis. Pubby welcomed the idea of having more to miss. People can create meaning even under dire circumstances.

I told Jake: "It comes down to what *you* think makes life meaningful. What would be sufficient, for you and for your children, to have led a meaningful life?"

How can you make life choices that serve you against the fraught backdrop of our warming world? Science tells us our remaining carbon budget and the carbon cost of different choices. But spending this budget wisely requires each of us to examine and identify our deeply held values and look at how to put them into practice to find meaning.

Beyond Happiness to Meaning

I want to be clear: As a scientist, I have extra authority on my subject of expertise, but no special insight into what makes a life worth living. Each of us has to figure out our own answer to that question and put it into practice for ourselves, a task for the ages. That said, you might have stopped asking yourself "What's the meaning of life?" after you went through that angsty poetry-writing and unfortunate beret-wearing phase in high school (or maybe that was just me), but you shouldn't. (Shouldn't stop asking the question, I mean. That beret was a serious fashion crime.)

I'll start by distinguishing a sense of meaning from a feeling of happiness. Long-term happiness depends on close relationships, a Harvard study found. More superficially, research shows that happiness is basically associated with people getting what they want and enjoying good health, material comfort, and lack of difficulties. A study of 12 million blogs found younger people tend to associate feeling good with excitement, reflecting their focus on the future, while older people associate positive feelings with peace and contentment, reflecting a greater focus on the present.

While both being able to look forward to the future and finding joy and gratitude in life's small daily moments are important, meaning is a much deeper, more transcendent and enduring experience than happiness. I believe meaning is not some fixed jewel, but rather something each of us creates from actions and relationships aligned with our personal core values: what you care about most, the principles you believe are important and right that guide your decisions. One way to identify your core values is to write down a list of the people you most admire and the qualities they personify. Studies show the values behind meaning are about giving rather than taking; they're associated with helping, taking care of, and even sacrificing on the part of others, and being part of a larger community or cause for good. This helps explain why parenting can feel like "all joy and no fun"; while surveys report that parents would be happier in the moment to watch TV instead of taking care of their children, they look back on caretaking as meaningful.

Lifting our gaze beyond only happiness to focus on meaning will help us dethrone some Exploitation Mindset values that don't fulfill us. In serving short-term desires, the Exploitation Mindset tends to prioritize convenience and efficiency over other values.

But convenience is not a core value that leads to meaning; it is not the point of life. We need to start a cultural shift toward a greater emphasis on what really gives meaning; to do so, we need to choose direction over speed.

The psychiatrist and Holocaust survivor Viktor Frankl describes three routes to meaning: through what you do (creating a work or accomplishing a goal), through what you feel (experiencing something or loving someone fully), or by the attitude you adopt toward the unavoidable suffering we will all encounter in life. Thus, deriving meaning from life isn't about avoiding suffering, as Pubby wrote, but about developing ways to face it with grace. Meaning can shine the brightest from the darkest circumstances, when a tragedy brings priorities into stark relief, or when we are inspired by stories of kindness and courage under inhumane conditions of war or terminal illness.

In our last exchange, Pubby dropped some final truth bombs, a gift from a dying friend to help me find my way. Pubby wrote me: "Reaching maturity is something like switching from thinking of your life as a first person narration to thinking of yourself as a character in a larger third person narrative. . . . Meaning comes from being able to make the story not about you, but about a larger narrative that goes on after you die. You want to be as big a character in that narrative as you can, and all the stuff that surrounds you, comes before and after you, is what grants your life meaning. Even if you lose . . . Anyone who has had the privileges and resources we've all had to craft a meaningful identity and didn't, well, boo on them—major opportunity missed. I understand 99.999% of humans who've existed primarily had to worry about where to sleep and what to eat for the least amount of work, and so I can't really blame anyone for living by those rules, but life has a lot more to offer, and you should take it."

Working to undo climate change provides a crucible to create meaning in our lives. It gives us a role in the greatest story ever being told, humanity's most epic test; a purpose in pursuing goals chosen to serve our core values and protect what matters most to us; and a way of mattering to others and the world. By Frankl's yardstick, our work toward stabilizing the climate could not be more clearly cut out for us. The climate crisis gives us ample opportunity to fully experience All the Feels ourselves and as part of our communities, and is turning up the heat and amplifying suffering, giving us a chance to cultivate our empathy and compassion while we help one another build resilience and carry on.

Today, nearly a quarter of Americans lack a strong sense of what makes their lives meaningful. To them, and to everyone else who already knows what matters most to you, I humbly offer a place on Team Climate. Our own lives matter; the lives of those we love matter; the lives of those we'll never meet, on the other side of the world, matter. Protecting these lives from harm, making them more resilient, giving them more options and capacity and kindness to face the future, matter. This is what our current climate crisis offers us: the chance to make our work (paid or volunteer) matter, in service of helping others; to find meaning in fully experiencing our relationships and experiences; even under difficult circumstances, to persevere, to maintain dignity and kindness. Fighting climate breakdown is not easy. It rarely feels like I'm really making a difference. But I deeply believe that it really matters, and from that I derive immense meaning.

Maximize Meaning, Minimize Carbon

Questioning what gives you meaning is a lifelong process. But how can you use your current answers to inform your life decisions, big and small?

We're used to making choices in terms of the limited resources of time and money, as the clock ticks relentlessly forward and our bank balances are unforgivingly finite. There are trade-offs between time and money: Sometimes we can buy services or products to save us time, like if we free our time for other pursuits by paying others to clean our house, care for our children or parents, or prepare our food.

There are of course limits to what money can buy. Even with outstanding healthcare, not all ailments can be cured; money does not *guarantee* more time. And while time is finite, it can to some extent be created—both by fully inhabiting the present to make use of the time we have now and by investing time to take care of our bodies and minds, to play the hand we're dealt by genetics and luck as long as possible. But ultimately, a ledger of how we spend our limited time and money reveals what we prioritized as mattering most.

I spend a lot of time thinking about a third limited and precious resource that we also spend in order to live out our values: carbon. Like time and money, how we spend our carbon says a lot about who we are and what's most important to us. We know we have a limited, very small carbon budget for all of humanity to share, now and forever. While different circumstances lead us all to have different reserves of time and money, we all have an equal share in the atmosphere. How we spend our personal carbon budget affects not only ourselves but everyone else's opportunities.

In a warming world, all of us urgently need to make choices that support what we genuinely need, to say yes to what nourishes us and gives us meaning and makes us feel more alive—and *only* those choices. Put another way: Let your focus on meaning, including your concern for the world, be the excuse you need to say no to the things in your life you don't truly burn for. A constant scramble to keep up with expectations is exhausting—and it often materially exhausts the environment as well.

Living ethically and responsibly in this era will ultimately involve some painful reconciliation, forcing us to come to terms with the very real costs of some of the overconsumption exalted under the Exploitation Mindset, including the jet-set getaways, shiny cars, and fat steaks that may still be filling your social media feeds. The least we can all do is stop giving even a nano-shit about the things we don't genuinely need, so that we can focus on using and enjoying what we already have.

A lot of changing your life for the sake of the environment means trying *way less hard,* not more. British researchers identified activities to live better and happier while consuming and emitting less; they found the best win-win leisure choices are local outdoor activities, reading, hobbies and games, music, and—lowest emitting of all—"sleep and rest." Science says: more naps!

So when what you're seeking is quality time with family in a beautiful natural setting, ask yourself if you really need to spend the time, money, and carbon it takes to fly halfway across the world to the latest "hidden" tropical beach—or is there another restful locale closer to home where you could get just as much fun and together time, and perhaps less stress? When you want your kid to have friends and enjoy moving around outside, is it worth the strain and pollution to drive her to soccer practice three towns over, or would she be just as happy messing around in the neighborhood creek with kids from down the street, while you enjoy a quiet morning at home? Learning to live well where we are is a regeneration skill to cultivate.

When I moved from the United States to Sweden, I was struck by how powerfully culture shapes what we value, expect, and strive after. Preparing for the move, I read a book about Swedish culture that described the concept of *lagom,* the idea that enough is not just "enough," but that a sufficient amount is the perfect, most satisfying thing. This just . . . did not compute. I simply could not wrap my American-indoctrinated mind around something that sounded like voluntary simplicity, in contrast to those constitutionally enshrined values, "bigger is better" and "more is more." Slowly, though, as I got used to my new surroundings, I realized that my life was not fundamentally improved by having two hundred brands of shampoo or yogurt to choose from at the store.

Similarly, I remember my first new friend in Sweden, Amanda, saying she always asked herself, "Do I need this?" before buying something new. This went way beyond sparking joy; how much of what I bought was based on an actual need? Asking myself this humbling question seriously reduced my stuff intake. When I do need something, I try to start with using or repurposing what I already have, then see what I can borrow or find secondhand.

The mindset shift of focusing on maximizing meaning in our decisions helps shift our focus toward deeper questions of what we truly value and what will contribute to regeneration, and away from pursuing the status markers of the Age of Exploitation. For example, it might shift our focus away from the question of which car to buy, to imagining what it would take to have a life (and a neighborhood, and a city, and a society) built around people instead of cars.

In sum, I think one measure of a life well lived is one that maximizes meaning and minimizes carbon. This way of thinking encourages us to prioritize the big stuff, the core values we can't compromise—relationships, health, helping others. Everything else can fill in the cracks, if there's space. It also makes us rethink what our real needs are and find creative, fulfilling ways to meet them with the least carbon possible. Looking through the lens of carbon can help order our priorities to focus on what really matters.

Work with Purpose

Recently I received a thoughtful e-mail from Elena, a sixteen-year-old from Detroit, under the subject line "Climate/Career Advice for a Teenager":

> I am upset about how many people in the world are suffering and will suffer in the future, especially because of climate change. . . . I was hoping you could give me some brief advice . . . about where to put my time and focus to help deal with crises such as poverty and climate change most efficiently. I receive different messages from my parents and the media that I should be convincing my community to reduce their carbon footprint, helping climate activists, or investing in my education so that I can be a better activist in the future. . . . If you were a sixteen year old with good science and social science skills right now, what would you tell yourself to study in college, and what career/activist route would you tell yourself to follow?

I told Elena I thought she was already asking the right questions that would help her figure out what she felt most drawn to and where she could have the most leverage. Change is urgently needed at the community, system, and individual levels she mentioned, and progress on any level reinforces and opens up new possibilities on the others. We need a diversity of messengers, and messages, to reach and resonate with different audiences. Rather than looking for one right answer from some external authority (the one and only, perfect pathway everyone must follow to solve climate change), the task for each of us is more about figuring out who we are and becoming more of that. Each of us can find ways to adapt our particular skills, talents, and preferences to our local context to work toward the principles of protecting people and nature, reducing harm at its source, and increasing resilience. We need to draw on the talents of everyone: architects to imagine and design the future, and builders to make it a reality; gardeners to grow food, caretakers to feed a hungry bunch, connectors to create community; storytellers to evoke emotion through art; communicators who inspire and persuade.

If you're in a position to choose or adjust your work, identify work you genuinely enjoy doing, that you're good at doing, and that you find meaningful. Doing what you think is fun, satisfying, and valuable will be the work you can sustain over time. (After I spoke about this idea of mine in a lecture, an attendee referred me to the Japanese concept of *ikigai*, the reason why you get up in the morning; it's often shown in the West as a Venn diagram of what you love, you are good at, can be paid for, and the world needs.)

I want to add an important caveat that I really needed to hear in my twenties: It takes time to develop the skills to be good at something, which you gain through doing a lot of (mostly not that great) work. So don't give up if you're not *yet* good at what you really want to do. (Thank you, *This American Life* host Ira Glass, for your reassuring advice that it was normal to suck for a long time while developing the skills to narrow the gap between your ambitions and the work you make!) I'll also add that I've found that gap never goes away entirely and that you have to keep examining if your current work still meets these *ikigai* criteria, as you change over time; my dad's advice is to never stop asking yourself what you want to be when you grow up.

For individuals pursuing careers and industrial policies alike, the key takeaway is that we need all jobs to be compatible with a stable climate. We need contributions from people from around the world, from all fields, industries, and perspectives, working on solving this problem from every angle. Fortunately, there's a growing movement of people refusing to work for or support companies that promote climate destruction, including thousands of students graduating from universities in France pledging only to work for companies "sufficiently committed to the ecological transition." My friend Rebecca has helped lead discussions on how the creative talents of people in the advertising industry could be directed away from "exacerbating our current climate crisis through promoting unsustainable consumption on behalf of our clients," and instead to harness "the very skills that have been used to shape the values, attitudes and behaviours of consumerism . . . to help shift society to more sustainable ways of living."

Making Decisions Guided by Meaning

Back to Jake's question about whether he and his wife should try for their second child. Part of the reason he contacted me was that Seth Wynes and I published a 2017 study quantifying the most high-impact personal climate actions. The most important choices for high-emitting individuals to quickly reduce our climate pollution were to live car-, meat-, and flight-free, as I'll talk about more in Part 3. But our data revealed something awkward. There was a fourth personal choice with an even bigger long-term effect on the climate: choosing not to have a(nother) kid. Put simply, in countries with high emissions rates, adding more people with today's consumption patterns adds a lot more carbon to the atmosphere—and their children will add more still.

Unlike driving SUVs and flying in planes, however, making the free decision of whether and how many children you want is a fundamental human right. The freedom of that decision is at the core of the humanity we need to protect from climate breakdown.

Not to mention this double-negative fact: Nobody's going to save the planet by *not* having kids. At current emissions rates, there are already enough of us overconsuming the carbon pie that it will be entirely gone before today's newborns can write cursive. It's today's emissions that need to be eliminated, so that people alive today and in the future can have a good life on Earth. For all these reasons, Seth and I tried so hard to be extremely careful with how we presented our findings to the media. *We tried so hard.*

But what Seth and I didn't quite understand yet is that all of us, all of our human brains, need to understand the facts of the world by placing those facts in the context of a bigger story. Moreover, when confronted with new facts, the idea of coming up with a new story doesn't typically occur to us: We reach for something off-the-rack, for the comfy, broken-in plotlines of ideology. And when a certain cadre of commentators saw our carefully presented facts, they reached for one of the most clichéd plotlines of all: ENVIROZ HATE BABIES.

While responsible outlets like *IFLScience* titled their coverage with the factual THESE FOUR LIFESTYLE CHANGES WILL DO MORE TO COMBAT CLIMATE CHANGE THAN ANYTHING ELSE, one blog accused our study of being part of the "culture of death" on the left, which was invoked to explain why "they push abortion, euthanasia, national healthcare (so government can decide who lives and who dies) and complete control over your life." (To be clear, our study mentions "carbon" forty-two times, and abortion, euthanasia, and healthcare exactly . . . zero times.) A commentator in the *National Review* admitted she had "no scientific credentials of any kind" but suggested that climate scientists should instead "drink the hemlock" and "commission a study on the climate impact of a liberal-environmentalist suicide pact." While not directly suggesting I kill myself, the English scholar turned climate opinion-haver Roy Scranton wrote in *The New York Times* that our study supported suicide. (Having experienced the devastation of suicide among my friends and family members, I hope it is abundantly clear that it did not.)

There were more disturbing responses, including some arguing that people like themselves should have children to save the planet. Personally, I reject the idea of using children as a means to an end, and I think it's deeply unfair to push adults' urgent responsibility to slash emissions now onto someone more vulnerable in the future. From a practical point of view, given that the remaining carbon budget is set to run out within the decade, future generations will be too late to meet the Paris goals. Even an incredibly gifted child is unlikely to carry out a global-scale job that humanity has been shirking for generations before they learn to ride a bike. I'm also concerned by arguments that frame having children as creating a larger army of "us" than there are of "them" to achieve a certain purpose, even a purpose as noble as climate stabilization. Such arguments quickly slide away from good intentions to raise more empathetic and resilient humans, and toward a supremacist favoring of only people who fit certain political, ethnic, or religious criteria, violating the core value of equality and the principle that all people have equal worth.

Personally, I applaud people who undertake parenthood as a vote of hope that the world is going to be a better place, along with accepting what author Keya Chatterjee describes as her parental responsibility to fight for climate stabilization to protect her son's future. Bucking the usual trend of household emissions rising after a new baby is born, she and her husband reduced their household climate footprint after they had their son, through choices like rooftop solar panels (which made a far bigger difference than diapering choice), staying in their house in the city rather than moving to a car-dependent home in the suburbs, and reducing their frequent flights to one flight a year.

In approaching their personal family decisions, I'd love to see more people actively examining if "one of your central goals in life is to procreate," as the bioethicist Travis Rieder puts it. My friend Michelle followed this advice in genuinely inquiring whether she wanted a baby. She spent a long weekend alone in a lovely grove of banksia, the iconic Australian shrubs that look like a cross between an artichoke and a feather duster. After two days, her mind had quieted, and she gently asked herself, "Do you want to be a mother?" She described feeling in her body a simple, primal, joyful, resounding answer: "Yes." I'm thrilled for her honoring this call. I'm equally thrilled for my friend Jessica, who is deeply fulfilled by her consciously chosen child-free life. Both of them are clear about their core values and are putting them into practice. (And of course, there is always an element of chance in the roll of the cosmic dice in the choice to have a child, as my friend Jennie says.)

Our challenge as human beings alive today is to redirect the flow of human institutions and culture away from the mental and material legacies of the fossil era. Needless to say, that does not involve dropping the paddle and letting the current just carry us along. Instead, while we confront scientific realities, we have to do some serious soul-searching about what constitutes a good life and how we make such a life possible for everyone, now and in the future. Please don't make your major life decisions just because you feel it's what People Like You Should Be Doing. Keep asking yourself what is essential for the story of your life to make sense to you, for your choices to align with what you value most, and how you can contribute to dismantling the values behind exploitation and cultivating those that lead to regeneration instead.

Chapter 6

Face Your Fears

Through Doom to Purpose,
by Way of All the Feels

I was in an academic seminar room, but I had to pinch myself to make sure I wasn't dreaming, or more specifically having a nightmare. A brilliant scholar I admire was describing things that I could not believe were serious policy proposals for 2020, not dystopic science fiction from 2200. Let me try to explain.

The proposal being floated was a geoengineering scheme known euphemistically as "solar radiation management," more specifically "stratospheric aerosol injection." The idea is to continuously shoot tiny mirrorlike particles into the high atmosphere to reflect incoming sunlight. After a few decades (!), proponents believe (!) temperatures will cool noticeably. The upshot: Prestigious scientists, including a panel from the US National Academy of Sciences, are seriously studying the suggestion that humans *fight the sun* instead of weaning ourselves off fossil fuels. I don't think I could have made up something more hatched from the Exploitation Mindset if I tried. Basically, this approach combines excessive faith in technology with fatalism about the possibility of human behavior and social change to conclude that humanity is incapable of getting our act together to address root causes, and we will have to settle for trying to mop up the mess we'll just keep making. Some proponents claim fighting the sun would be a way to "buy more time"—as if putting off the inevitable clean-energy transition until it's more desperate will make it easier or smoother.

In reality, going down this geoengineering path would be a permanent commitment; if we ever stopped fighting the sun, the full brunt of the carbon pollution still accumulating in the atmosphere would be felt in "termination shock," potentially rocketing up temperatures several degrees over a few years. And that carbon building up in the atmosphere condemns the oceans to keep acidifying toward death. Plus the sky-mirror particles themselves cause problems, like potentially shutting down the circulation that drives the Asian monsoon, which would disrupt climate patterns that billions of people depend upon for sustenance. It would feel like the sci-fi nightmare it is, because the sky would no longer be blue, but a dead, washed-out white. Oh, and we would never see the stars again.

This is where the Exploitation Mindset leads, and this is not the sky I want us to make. We need to find better ways of honestly facing and starting to repair the climate chaos we've made so far. To do so, we will have to acknowledge the difficulty of perseverance in the face of some very tough odds, but then find ways to do it anyway, cultivating our own resilience and the support of our communities. To get on the path to regeneration, we're going to have to face All the Climate Feels, including our fears of the worst, and the ever-present temptation to just give up.

OK Doomer

With desperate proposals like geoengineering being discussed by Serious People, it's hard not to despair, or turn into a climate doomer who concludes nothing we do matters. But cynicism is a form of denial—of ignoring the very real things you could be doing to salvage what you can. It's only a bit further down a slippery slope to fatalism about the resilience of nature, or human nature, that leads you to decree that it's too late for the climate and for human civilization, so bring on the bunkers and the bourbon. I think this doomerism is wrong. Innocent, young, and vulnerable people and species don't get to give up or build bunkers as they live with mounting climate hazards, and neither should the privileged.

I take climate fatalism as a litmus test of your views on human nature, not the nature of biophysical reality. There are lots of studies showing it's technically possible to solve climate change. As one author of the IPCC *Global Warming of 1.5°C* report put it, "Limiting warming to 1.5°C is possible within the laws of chemistry and physics but doing so would require unprecedented changes," which the report authors further clarify as "rapid, far-reaching, and unprecedented changes in all aspects of society." (I'll get into more details on what those changes look like in Part 3.)

Is humanity actually capable of rising to this challenge and making the necessary, unprecedented changes? No one knows! Social science and history can give insight from past experience, but there is no parallel experimental or historical precedent to draw upon, although certainly humanity has done a lot of epic, amazing things that seemed impossible until they were done. People who think humanity is too lazy and selfish to bother, that everyone will stick their head in the sand and wall themselves off and step over others to get the last loaf of bread, might be right. But I want to be on the team that's trying to prove them wrong.

The Struggle Is Real

I don't think the doomers should be determining the future of our civilization and our planet, but I empathize with their cynicism; I really do. I don't want to belittle the temptation to give up in the midst of this mess, because this work is really hard, and I struggle with this temptation myself. Humans have set in motion a planetary-scale warming experiment with impacts that will last for millennia; the headlines about the latest ice-sheet collapse or coral bleaching only hint at the scope of the destabilization we're fighting against. There is so much we have to do, and so little time to do it. And I, just like you, am just one person.

Even though I spend most of my time working on climate, if I'm honest, most days I don't have any sense that what I'm doing makes any difference in the world, to the climate, or to other people. What if it is too late for anyone to do anything to make any difference? What if it is already too late for 1.5°C or 2°C because of tipping points lurking just around the corner? Will all my efforts have been a total waste of time? Should I have started that garage winery instead?

On a particularly bad day recently, when the headlines were especially grim and I felt like nothing I'm doing is making any difference at all, I came across a piece on climate activism by Jim Shultz. (I think I googled something like "climate change when to give up?" Please don't judge.) Shultz writes that there are no ways to know what impact we actually have, and no guarantees that what we do will make a difference: "So we guess, and there are two different ways we can guess wrong. The first is to overestimate our power to change what's coming and to give people . . . 'false hope.' . . . The second is to underestimate what is possible, to believe that we are less powerful than we actually are and to do less than we can. That's the wrong guess that worries me more. Faced with a choice between disappointment or failing to do all that is possible, I don't find the decision a hard one to make."

You Don't Have to Be Perfect; You Just Have to Be Brave

Many people would like climate scientists to give them hope. This is complicated, because it is not honest to say that everything is going to be fine. I think it's more helpful to think about taking action as a spark for hope, not the other way around. The Swedish climate activist Greta Thunberg agrees: "Once we start to act, hope is everywhere. So instead of looking for hope, look for action. Then, and only then, hope will come." I take comfort in Rebecca Solnit's description of hope: "Hope is not a lottery ticket you can sit on the sofa and clutch, feeling lucky. It is an axe you break down doors with in an emergency."

The temperature goals are truly, critically important; they have real meaning and are worth fighting for. But what if we miss 1.5°C, and then we miss 2°C? As the climate scientist Kate Marvel says, "Climate change isn't a cliff we fall off, but a slope we slide down." There is no bottom to the slope; we will keep sliding until we put on the carbon brakes. Every inch of the slide we don't cede is a victory.

How can we gather the courage to start honestly facing climate change and taking action? As the Heath brothers write about making change happen, identity can powerfully shape our decisions, motivating action if we aspire to be the kind of person who would make a given change. This invites us to ask ourselves three questions. First, what kind of person am I? Second, what kind of situation is this? And third, what would a person like me do in this situation?

I'd like the answer to the first question to be: I am someone who cares for others, who takes responsibility for my actions, who does the right thing even when people around me might not. My friend Lucy tells her daughter, "You don't have to be perfect; you just have to be brave."

As for the situation: Climate change is an emergency, and it's time we started treating it like one. Even when many people aren't yet acting like it is, even when powerful entrenched interests defend a status quo leading to a scorched planet, even when unhelpful voices get amplified and the silent majority who care don't dare to speak up or aren't sure what to say.

So what does a good-hearted person do in an emergency? They look for ways to help, and they pitch in where they can. They assess their assets and liabilities, talents and communities, and figure out how to mobilize them toward protecting people and nature, reducing harm at its source, and increasing resilience.

The Five Stages of Radical Climate Acceptance

From my own experience, partly drawing on inspiration from the work of colleagues from psychology, I believe the messy, ongoing work of facing and ultimately coming to terms with climate reality can be understood in five nonlinear stages: Ignorance, Avoidance, Doom, All the Feels, and Purpose. I believe that recognizing these stages, being aware of how you progress through and between them, and supporting others in doing the same, are key to transforming your experience of facing climate change from feeling helpless and overwhelmed to feeling empowered as an agent of change and part of a broader community. Some people equate accepting with giving up. But I'm talking about a different kind of acceptance: viewing reality as it is, without distortions or judgment, to change what you can and accept what you can't. Remember, pain times lack of acceptance equals suffering.

Let's start with **Stage 1: Ignorance**. This is the state we all start in, and we remain there until some conversation or headline first enters our previously blank climate consciousness. I think the first time I heard about climate change must have been watching a bizarre 1990 Earth Day special, featuring every celebrity of the day from Robin Williams to the Fresh Prince of Bel-Air to Kermit the Frog. Dr. Carl Sagan explained to a concerned crowd outside the hospital where "Mother Earth" (played by Bette Midler) was in critical condition: "In the air, there is a kind of invisible blanket of gases that keeps the heat in. . . . That's called the greenhouse effect. . . . It comes from burning fossil fuels: coal, oil, gas. . . . Most [scientists] think there will be significant warming by sometime in the next century." By 1997 I was taking ecology and geology classes at Stanford from professors who studied climate change and hired me as a research assistant on teams studying its unfolding impacts.

But for most people, the end of climate ignorance was much more recent. Americans interviewed in 2019 reported hearing more about global warming in the media over time, but still only half hear about it once a month or more. It's not helpful that the media, governments, and schools rarely talk about climate, and when they do, they often fail to provide the most accurate or relevant information and miss the links between unfolding stories and climate.

Nonetheless, after noticing or becoming aware of the problem, it can still be easier to avoid than to face. It's easy to scroll past the headlines announcing the latest record heatwave, hurricane, wildfire, melt, extinction, suffering. **Stage 2: Avoidance** is a mechanism to cope with feeling overwhelmed. It can be hard to reconcile our understanding of our lives in the context of the climate crisis. When a Yale survey asked Americans how often they hear people they know talking about global warming, the most common answer was "never," and less than a quarter heard these conversations more than once a month. The gulf between the seriousness of the climate crisis and a comfortable daily routine that doesn't acknowledge it is a very anxious and uncomfortable place to inhabit. This cognitive dissonance caused by the gap between your knowledge and actions may lead to the tendency to deny or diminish the uncomfortable information to avoid having to challenge your identity, beliefs, or practices, and to justify continuation of the status quo, especially if it benefits you.

But real problems get so, so much worse when you're hiding from them or pretending they're not there at all. One day, it becomes less uncomfortable to face the cognitive dissonance than to continue avoiding it. Maybe it's gradual, as you notice others around you increasingly taking climate action. Maybe it's a specific event affecting you or someone you love: a stronger hurricane, a more destructive fire, higher flooding. Or just a creepily warm day on Christmas, or spring flowers blooming on Valentine's Day, that makes you think, "This can't be right." This is where your googling to get more knowledge can take you down the black hole of doomsday feelings that Jake described in the previous chapter. The future can feel like a waterfall you're about to go over, or a wall you're about to run into; it's impossible to imagine what's on the other side. You may feel scared, paralyzed, lonely, or hopeless. This is **Stage 3: Doom**, and it's rock bottom. If you get stuck here, you may wallow in despair, or wall yourself off in anger, or else pretend you don't care and take self-righteous solace as a climate doomer to absolve yourself of all these uncomfortable feelings.

Good news: There is a way out of Doom! Bad news: It's *through* all those uncomfortable, "unfixable" emotions like grief, anger, and fear. These feelings are normal and healthy responses to the unjust and stark reality of the climate crisis. To navigate **Stage 4: All the Feels**, you need to find ways to acknowledge and tolerate them, to notice as they come and, inevitably, go. The kinds of practices that make you feel good and build up your physical and mental health and social connections and support (walks, journaling, sleep hygiene, healthy eating, mindfulness, therapy, music, wine and laughter and tears with BFFs) are what help you cope with this stage. They allow you to fully experience the present, and they keep you going for the long haul.

Honoring the importance and validity of feelings was something I learned the way most people do: the school of time and heartbreak. Saying the hard truths out loud can make me braver and stronger and wiser and kinder. If there's no knowledge or feeling to avoid, there's much less to fear. Even if the challenges in the external situation are the same, noting and sitting with them, rather than avoiding them or trying to change my reactions, feels like watching a thunderstorm through the window from a safe, cozy house, rather than being outside, drenched and lashed by the storm.

Relationships of care and social support are so much of what life's about. Cultivating face-to-face relationships with our neighbors and communities is absolutely essential for our personal as well as community resilience. These relationships also inspire us forward; as my students concluded their 2019 graduation manifesto: "Supportive communities can help us to be willing to change." Getting together to adopt a local park, advocate for safer streets, or tend a shared garden strengthens bonds that, in good times, give you folks to share power tools, childcare, and extra zucchini with. Crisis makes clear just how intimately our daily lives are linked, and these decentralized, self-organized, local networks are critical to helping us help one another get through.

Building and strengthening community lets you pick others up when they're down; they do the same when you need them. Even through the hardest times, until it actually ends, life will recalibrate on the other side of every heartbreaking loss. There will still be significance and beauty and kindness in your life, alongside deep sadness and grief and rage. And there will be important triumphs. Together, you find ways to go on.

This is when you reach **Stage 5: Purpose**. Purpose is the North Star you're aiming for, the point toward which you are striving. Your core values set the direction you think is important and right, pointing you toward purpose. If goals are set as specific milestones along the path between values and purpose, then over time, achieving your goals adds up to serving your purpose. Remember that pursuing purpose is part of achieving meaning. Today, 40 percent of Americans do not agree that their lives have a clear sense of purpose, so you may have to start with clarifying your core values, as we discussed in the previous chapter. You also need to assess yourself, your talents and strengths, to identify the piece of the puzzle you can help solve. (If you're stuck, ask family and friends when they've seen you at your most alive, and look for the common threads in their stories.)

Then look for ways to put your strengths to work for meaningful climate action in line with your values. How could you apply the principles of the Regeneration Mindset to protect people and nature, reduce climate harm, and increase climate resilience to your daily life at home, at work, in your community, your nation? Draw on your values and your strengths to form specific goals that make sense for you: Cut your household's carbon footprint 50 percent in the next three years. Build a community garden. Join a group pushing politicians to enact substantive climate policy.

Some of the things you try work, and some of them don't; you learn from both, and you keep going. You're surprised how quickly you adjust to a new normal, and you appreciate the unexpected upsides to new practices. When you struggle through the stages of climate acceptance and start putting your values into action in pursuit of your personal purpose, you inspire others to do the same, growing the community and its resilience.

One Foot in Front of the Other

The race to stabilize the climate is one I'm going to be running the rest of my life; its urgency notwithstanding, I know it's a marathon. To guide me on this journey and prepare for the road ahead, I find some useful lessons from the actual marathon I completed with my friend Jennifer in May 2011: Gather your peeps; set goals with your purpose in mind; the challenge is 90 percent mental; bring your favorite snacks; wear comfortable shoes; don't give up.

The night before the race, Jennifer and I had a pasta dinner in a tent with a festive atmosphere, with dogs sitting on chairs and local musicians performing. Everyone except the table full of unsmiling, lean professional runners was downing wine. The next morning, at the starting line, we stood bouncing around nervously with a lot of fit fifty-something French men wearing exceedingly short shorts. They all seemed to be named Laurent. The gun went off, and we started shuffling up a hill. For the next five-plus hours, we traced the Normandy coastline. We were joining centuries of pilgrims who had come this way before us.

Our destination was Mont-Saint-Michel, the towering abbey keeping guard over the salt flats on an island a kilometer offshore. How many people worked over how many centuries to build this exquisite monument? How many added their piece without ever knowing all the others who came before or after, or even knowing those who were working toward the same goal during the same years? How many contributed without living to see it finished? Their work still stands. Their island withstood sieges through the Hundred Years' War. It inspired the young Joan of Arc.

From the starting line, Mont-Saint-Michel was *barely* visible as a tiny triangle on the horizon. It looked impossibly small and distant. I couldn't imagine that my own legs would be all that would carry me there. The journey seemed far too long and uncertain. So much could go wrong.

I reminded myself of my last practice run a few days earlier. I set out filled with nerves about the prospect of running a marathon; it sounded hard. I thought of my training runs and wondered if they were enough. After I laced up my shoes and set off, I realized that actually, my task was simple, if not easy. All I had to do was put one foot in front of the other and not stop until I was done.

The day was uncomfortably warm and humid for running. I kept thinking I had to pee, even though I didn't. They had sugar cubes at the refreshment stations, when I craved salt.

We were among the slowest runners. Actually, if I'm honest, I'm pretty sure we were the very slowest. (*Chapeau,* Laurents!) The course officials started closing the course behind us, then alongside us. At one point I had to argue with them in my two-hundred-word French vocabulary to let us keep going. My goals when I decided to run this race were to not get hurt, to have fun, and to finish. I wasn't sure if I was currently achieving either of the first two, but at this point I'd be damned if I didn't reach the last one, even though giving up felt like a very real option at every step until the last.

Jennifer and I treated the marathon as a fundraiser for a blood cancer charity and asked for donations from our families and friends. More than fifty people had generously supported me. I had asked Jennifer to write the names of my donors on my right arm in Sharpie. I wanted to carry them with me on my run. The day before the race, sitting on a dock in the beautiful sandcastle town of Saint-Malo, she spent an hour carefully lettering their names. They filled my whole arm.

Around mile twenty, I started to hallucinate a bit. I didn't really feel like myself. It felt like my brain was hovering a few inches above my body, which was good, because my body was becoming an uncomfortable place to inhabit. It was starting to complain insistently and was becoming very interested in the idea of no longer continuing to put one foot in front of the other. I acknowledged this desire and did not give in to it. I kept going anyway.

I thought about what I was doing now, choosing to temporarily suffer, when it would be much more comfortable to just stop. It actually felt amazing to have the option to suffer, knowing it was a choice I pursued for a purpose I had decided. I couldn't stop now.

I felt the solidarity and support from the team I was carrying with me on my arm. It occurred to me how many of them were suffering from much bigger causes, causes they had not chosen and could not escape, could not choose to turn off after a few hours. A kind colleague had recently suffered the death of his teenage son, a healthy kid killed by a previously undetected heart defect during a high school football practice. I wrote him a condolence card with a photograph of the ocean I had taken from a boat, no shore in sight, just waves. A few months later, his house burned to the ground in a wildfire. He and his wife and dogs had escaped, but they lost everything. When I next saw him, I gave him a hug and said, "I'm so sorry about your house." He shrugged. "Thanks. It's just stuff."

I thought of how much suffering there was in the world, part of the shared human condition. I thought about the human capacity to endure, to keep going, to just keep putting one foot in front of the other in difficult circumstances, against the odds. Even in situations that seemed hopeless or where defeat seemed certain, some people kept going anyway. They persevered. And sometimes, they prevailed.

At that moment, I felt connected to my friends that I carried with me on my arm, and to the human spirit, tough and fragile and resilient. Enduring. It was a little leap of elation. In the face of so much courage, so many who managed to continue under much more difficult circumstances, I could certainly keep putting one foot in front of the other. I did. I do.

Chapter 7

Get Angry

Being Cassandra in a Post-Truth World

E-mail forwards from my dad usually run toward goofy animal clips that make me smile. But a recent one caused my head to explode.

It was an essay from a think-tank economist (note: *not* a scientist) named Irwin Stelzer arguing for a carbon tax. I'll come back to my healthy skepticism of carbon taxes in Chapter 11, but what struck me in this essay was the author's opening claim, which essentially proclaimed that it was impossible to ever know any facts.

Stelzer asserted, "We can't be certain that the globe is warming, and we can't be certain that it isn't," since "both sides" in US politics have strong ideological reasons for taking their positions. He asked his readers to "ask yourself whether the globe is warming, and you would have to confess that you do not know with certainty."

I started writing a reply to my dad. I typed and retyped, until I found myself writing the sentence "Do you agree there exists an objective reality separate from political ideology?" at which point I gave up and curled up on my bed to cry in rage instead.

Of course politicians have varying ideological reasons to

support their views; that is their job. But how have we gotten to the point where I am trying to argue with my father about whether politicians write not just the laws of the land but also the laws of physics? Why would Stelzer or anyone else think that assembling and interpreting records of observations about the material world— in other words, *science*—is the job of politicians?

Science is the job of . . . scientists. Irwin Stelzer, if you want to know if it's warming, please, for the love of all that is good: Ask a scientist. Ask the thousands of records and observations scientists have carefully curated across centuries and continents. Read the litany of statements from scientific societies. They all converge on the answer: Yes.

Or don't ask anyone at all. Don't listen to politicians or scientists. The beauty of the scientific method is that it's available to everyone. All you need is your own physical senses and a simple thermometer to make repeated observations (over thirty years for the case of climate averages), and logical reasoning to interpret the results. If anyone, anywhere follows the same method, the results will be the same. A thermometer is not liberal or conservative, as Katharine Hayhoe says.

Unlike faith (a personal belief that's impossible to prove right or wrong with evidence) or opinion (where, unlike facts, every citizen is entitled to their own), scientific truth does not depend on the instrument or observer. The material world exists independent of what humans think or feel or do. As my favorite protest sign from a 2016 rally of earth scientists read, ICE HAS NO AGENDA. IT JUST MELTS.

Fact is not an opinion you may believe to varying degrees; it's a truth you either know or you don't. Whether or not the Earth is warming is a factual and eminently knowable question, and the answer is yes. It is unambiguously, definitely warming, whether presidents or pundits believe or think or say so or not. If any of them had asked a qualified expert, or used their own damn eyes, they would know that with certainty.

We're Sure

Research is basically asking and answering questions, as my PhD advisor Chris Field told me. Following the scientific process, researchers formulate precise questions and use careful and conservative analysis, with a high burden of proof, to methodically test and eliminate all other possible answers to conclude that theirs is correct. To publish these results in the most reliable source of scientific evidence, a peer-reviewed journal, you submit your work to a journal editor, who recruits independent experts to scrutinize and either reject or improve it through the often-brutal process of peer review. If it's not rejected outright, you revise (and revise) until the reviewers are satisfied with the accuracy, rigor, and clarity of your work; then you share your results with the world, available for anyone to examine, replicate, and critique. Over time, evidence accumulates to answer a question with increasing certainty. Scientific consensus is the point at which scientists give up arguing about a question of fact because, as the science education website Skeptical Science puts it, "the sheer weight of consistent evidence is too compelling, the tide too strong to swim against any longer." Through this process, slowly, incrementally, the truth will win out.

Science has understood the basic physics of climate change for a *long* time. In 1896, the Swede Svante Arrhenius (a distant relative of Greta Thunberg) suggested that burning coal could cause global warming. Arrhenius spent a year making tedious calculations by hand, including using an estimate of the temperature of the moon based on moonbeams. His final estimate for the temperature increase expected under doubled atmospheric carbon concentration is in the same range as the most sophisticated one today.

By following the steady process of science over time, scientists have reached an overwhelming, complete consensus, backed by national academies, government agencies, and scientific societies: The climate is warming, and it is due to "human activities." Those activities are burning fossil fuels for electricity (25 percent of greenhouse heating), industry (21 percent), transportation (14 percent), buildings (6 percent), and other energy (10 percent), as well as 24 percent from how humans use land for food and forestry.

Since 2004, studies of studies analyzing the peer-reviewed literature have been used to assess the evidence for human-caused warming. The first such study, in 2004, found 0 studies out of 928 that disagreed that humans were warming the climate; a 2016 update applying this method found 99.94 percent consensus. Using many more studies and the stronger criteria that they must explicitly state humans are the primary cause of recent global warming, John Cook and colleagues found a consensus of 97.1 percent. This consensus was reaffirmed in 2016 in a study of the studies of studies(!). Importantly, the greater the relevant expertise of the authors, the more strongly they agreed humans are warming the climate. In short: Experts are sure that it's warming and it's us.

This overwhelming scientific consensus is critically important, because research shows that understanding the scientific consensus on human-caused warming is a "gateway belief" that leads to stronger support of meaningful climate policy. However, only about half of Americans currently know about the overwhelming scientific consensus on human-caused warming, and consensus is rarely discussed. Our recent study showed that scientific consensus was the least-covered topic in Canadian high school climate curricula.

So why is the overwhelmingly established scientific fact that humans are warming the climate not universal common knowledge? Where has this anti-science view of climate change come from, and where is it leading us?

Big Oil Lied Big

I often show this quotation in my lectures and ask students to guess who said it, and when: "There is still time to save the world's peoples from the catastrophic consequences of pollution, but time is running out. . . . Carbon dioxide is being added to the Earth's atmosphere by the burning of coal, oil, and natural gas at such a rate . . . to cause marked changes in climate."

The most common guess is Al Gore, in 2006. Barack Obama, in 2014, is also a popular contender.

The correct answer is the president of the American Petroleum Institute, at the organization's annual meeting. Heartbreakingly, agonizingly, terrifyingly: It was the annual meeting of 1965.

To understand how it has come to pass that we are already in such a very deep carbon hole, there is a painful and infuriating truth that can't be ignored: For many decades, organized disinformation campaigns by a small group of fossil fuel executives and their lackeys have aimed to deliberately sow public doubt about scientific conclusions by trying to silence, delegitimize, and discredit scientific knowledge, the scientific process, and scientists ourselves.

For my entire career, and for more than my entire life, far too many business, government, media, and cultural leaders have either ignored or actively denied climate reality, and either marginalized or threatened the scientists who gave it voice. Meanwhile, amid these decades of deliberately manufactured doubt and delay, fossil fuel emissions have tripled since the industry's own 1965 warning of catastrophic consequences.

I am enraged and despondent that lobbyists have spent decades spewing, and the media spent decades disseminating, misleading stories aiming to cast doubt on the settled science that the globe is warming because people are burning fossil fuels. The seeds of this false doubt were sown by people acting in deliberate bad faith, were propagated both knowingly and unknowingly, and have now rotted into genuine but unfounded public uncertainty and confusion.

Historians Naomi Oreskes and Erik Conway showed in their 2010 book, *Merchants of Doubt,* how the same strategies and scientists were used by Big Tobacco and Big Oil to sow false doubt about the science. A 2017 study by Geoffrey Supran and Naomi Oreskes showed that four-fifths of both scientific studies and internal documents from ExxonMobil from 1977 to 2014 supported the consensus that climate warming is "real and human-caused." However, at the same time, Exxon was paying for "advertorials" in high-profile media like *The New York Times,* 81 percent of which expressed doubt about these established facts. Many of the advertorials claimed the science was too uncertain to act, or that it would be rash and too expensive to make a transition off of fossil fuels. The authors conclude, "ExxonMobil misled the public."

The business plan of fossil fuel companies remains unchanged: Continue and expand carbon extraction and burning, which a 2019 InfluenceMap analysis showed is where 97 percent of capital investments for the five largest oil majors still goes. A 2020 Carbon Tracker report analyzed oil majors' business plans and confirmed that they don't meet even the 2°C goal, despite their claims. Yes, right now, fossil fuels give us hot showers and cold beers. But there are ways to deliver the services we need (and even cold beers) that don't cook the planet.

You'd never know how much Big Oil is doubling down on fossils from looking at their ads and social media, which are filled with misleading campaigns implying they're transitioning to clean energy. Oil majors spend 29 percent of their lobbying and advertising budgets to promote themselves as low-carbon leaders, a proportion almost ten times higher than the 3 percent of capital budgets they're actually investing in clean energy. A particularly egregious example comes from Exxon's goal to produce biofuels from algae. Despite their splashy ad campaign, their algae plan by 2025 adds up to just 0.2 percent of their current fossil-based capacity. Meanwhile, a CDP analysis showed that the top one hundred oil companies are responsible for 71 percent of all industrial greenhouse gas emissions (those excluding land use) since 1988.

Lives and species are being lost because of fossil fuel companies. I truly do not understand how oil executives can sleep at night or look their kids in the eyes. I am filled with rage every time new evidence of their harm, deceit, and hypocrisy comes to light.

Can Reality Survive in a Post-Truth World?

Worst of all: The dirty tricks of fossil interests *work*. As a result of their disinformation campaign, there is still significant doubt in particularly the American mind about the scientific reality of human-caused climate change. This has translated into weakened support, increased apathy, and what Alex Steffen calls "predatory delay" toward climate action. A 2018 Yale survey found that, fed a steady diet of messages of doubt in scientific consensus, only one in five Americans correctly understands that basically all climate scientists agree that human-caused global warming is happening.

Undeterred by the overwhelming weight of scientific consensus, the hacks are still trying to drive public discourse. Amber Kerr of UC Davis analyzed a 2019 letter signed by "500 scientists" claiming "there is no climate emergency" and found that just 2.8 percent of the signatories identified themselves as climate experts. The largest group, 21 percent, were engineers, many involved directly or indirectly with fossil fuel extraction; 19 percent were geologists, and a further 11 percent were nonscientists, including business executives and lobbyists. As Upton Sinclair wrote, "It is difficult to get a man to understand something when his salary depends upon his not understanding it."

The gaslighting of reality invades language in a positively Orwellian turn, when scientists can't even talk about what we're actually talking about. My scientific colleagues in the US federal government were ordered in 2017 to avoid using the phrase "climate change" to talk about their work; the Environmental Protection Agency's website scrubbed mention of climate or fossil fuels or even "science"; Floridian civil servants must talk about "nuisance flooding" instead of sea level rise. After using fake words to convey a verboten meaning, will we eventually forget the real name of what we're witnessing? Will withholding a name for what we're seeing help us ignore that it's really happening?

As a scientist, I find this anti-truth campaign deeply distressing. In my own experience, it's no fun to face the personal and professional attacks from the chorus of trolls who feed on and amplify misinformation (though at least I'm in excellent company with every other female scientist I know). I'm angry and concerned on behalf of friends and colleagues who have received harassment—up to and including death threats—to try to silence their climate research findings.

But I think the biggest effect of climate change denial is more existential. It is deeply disturbing to worry that facts don't matter anymore. It's exhausting and infuriating to be a scientist in a post-truth world. It is a struggle to sustain the belief that truth still exists in the face of blatant denial of not just specific evidence but of the existence of independently observable evidence as a way to determine veracity. It is tough to get out of bed, much less to speak publicly about science, knowing some people do not believe scientific expertise qualifies scientists more than politicians to speak for, well, science. It is destabilizing to society when spin can overcome evidence, to the point where intelligent people can find themselves arguing over the existence of physical laws instead of having a debate about political preferences.

Worryingly, party politics has become a marker for belief in scientific reality. A 2016 study across fifty-six countries found the strongest correlate of climate change "belief" was political affiliation, with liberal voters "more likely to believe in climate change." But the United States is a global outlier in this field. Unique among major political parties in the democratic world, contesting the reality of the warming climate and its seriousness has become part of the Republican Party line (forcing increasing divisions within the party). Science has been politically weaponized in the United States, so that climate change is now the issue with the single strongest polarization between Democrats and Republicans—stronger than gun control or abortion. For those ideological issues, people disagree over which values should be prioritized to determine what is right or wrong. But with climate change, partly due to "solution aversion" of interventions perceived to rely on big government, people can refuse to accept externally verifiable facts. This means that more facts cannot change the hearts and minds of people who deny the fact that the climate is warming.

Being Cassandra

I've felt the constraints imposed by politics and self-censorship on what I can say as a scientist since the first study I coauthored was published in 2004. In my first year of my PhD at Stanford, my advisor invited me to join him and a team of highly qualified senior scientists to contrast the future of health, water, and agriculture in California under two different scenarios: one where we reduce emissions, and business as usual where we don't.

My analysis showed that continuing to produce high-quality wine might be impossible in most of California under unchecked global warming. I worried about the wine industry response to our study, which is to say, I worried about how my family and our neighbors and friends would respond, because the wine industry drives two-thirds of the economy and employs nearly a third of people in Sonoma County, where I grew up, and more in Napa County, where I went to high school. Our results were bad news for locked-in investments, for tourism, for the price wine buyers might be willing to pay if they perceive a region is losing its terroir. Those are people's livelihoods and family businesses on the line. I felt great responsibility to communicate my findings carefully, to make sure I made clear the assumptions and uncertainties, and most of all to convey the monumental choice we still had to get on a low-emissions pathway and avoid a lot of harm.

Our study received AP coverage reprinted in *The New York Times,* which opened: "Global warming could cause dramatically hotter summers and a depleted snowpack in California, leading to a sharp increase in heat-related deaths and jeopardizing the water supply, according to a study released Monday. The report is substantially more pessimistic than previous projections, and was dismissed by one expert [*sic*] as 'another piece of climate alarmism . . . designed to paint a very frightening picture.'"

Unfortunately, this kind of journalism giving equal weight to "both sides" of a scientific issue, with actual scientists on one side and special interests on the other, used to be common even in highly reputable media. In this case, one side was a team of nineteen scientists from the top universities and national laboratories in the country, who undertook a year of analysis that was published in a prestigious scientific journal after rigorous peer review. The other side was the spontaneous opinion of one lobbyist from the Competitive Enterprise Institute, the Koch-funded group with long ties to tobacco and climate disinformation campaigns, who more recently led opposition to the Paris Agreement. These sides are not and should not be regarded as equally scientifically credible. Depressingly, my experience fits a larger pattern. Rachel Wetts of Brown University analyzed more than seventeen hundred press releases from 1985 to 2014, finding major US media outlets were more likely to cover business interests than science, and twice as likely to cover statements opposing climate action as those favoring it.

Climate scientists reporting our alarming findings have faced derision for being "alarmist," with its irresponsible and unscientific implications, for decades. I remember hearing a lecture in grad school from Naomi Oreskes, the Harvard historian, recounting a story where she gave a fact-filled climate lecture to a public audience. After her talk, a woman stood up and said, "What you're saying is alarming. But you don't sound alarmed." The critique stuck with her, and Oreskes realized that the dispassionate language and demeanor of science were failing to convey the true magnitude of the alarm. She urged scientists to put more passion into our communication, so that our delivery would convey the urgency of our findings. It's not being an alarmist if you are deeply alarmed by distressing findings.

Still, it's uncomfortable to sound the alarm. "Alarmism" has haunted me long since. In 2017, I was part of a Reddit Ask Me Anything on climate with Seth Wynes. A Redditor wrote to ask if we thought the threshold of 400 ppm of atmospheric CO_2 we'd recently blown past was "a genuine tipping point or just symbolic," and whether this meant that climate change was irreversible. I wrote that I was "alarmed that humanity had crossed a threshold for the first time in 800,000 years." Then I sat there and debated with myself about my use of the word "alarmed." The contrarians' accusation of being an "alarmist" rang in my ears. Was I just playing into their hands? I wrote and rewrote my response and discussed it with Seth over our private chat. Finally, I hit send with "extremely worrisome." Which one was right? How do I put a statistical value on a feeling, informed by decades of training, but ultimately interpreted through the tightening of my gut?

Being a climate scientist feels like living through a slow-motion nightmare I can't wake up from, because the nightmare is happening in the real world. It is being Cassandra, destined to speak the truth without being believed, until it starts to feel hopeless and useless to speak a truth it seems no one wants to hear, a truth that gets drowned out by clever PR campaigns and talking points designed to suppress or diminish it. Telling the truth is not enough.

I think back to a set of studies I collaborated on with David Lobell and other colleagues at Stanford around fifteen years ago, where we used temperature and rainfall data from the previous twenty-four years to understand how climate drove the yields of many of California's iconic crops, including wine and table grapes, oranges, almonds, walnuts, and avocados. Looking to the future, we saw that regardless of the emissions scenario, we could expect that with the warming we had already locked in, sometime soon, my home state would be in a "novel climate regime" where for growers, "relying on past experience may be a poor guide when climate warming exceeds the thresholds of natural variability, as projected beginning in 2020 in California."

At the time, our analysis felt abstract and distant; 2020 was a point on a graph. Now that date is here, and we see that in

California, and in the rest of the world, we already inhabit un-charted terrain. Global warming has already redrawn the maps. This diminishes the value of generations of farming know-how and tradition and best practices gained from experience and re-search in the old conditions. Now those expectations for when rain would fall and how heavily, when to plant crops and how to prepare for frosts, are no longer valid. It's as though the earth has moved under our feet, and we've woken up in a new country, one that looked at first completely familiar but no longer followed the same rules we'd always grown to expect based on our expe-rience.

I don't feel at all triumphant to have been right about devastat-ing impacts; I desperately wish all the climate science were wrong. I am crushed every time I read a quote from a scientist saying the latest impacts are exactly as predicted, but "decades ahead of even the most pessimistic climate models," as Simon Lewis told *The Guardian* in March 2020 about his study showing the Amazon rain forest is losing its ability to take up carbon. I take some sci-entific pride, but zero personal satisfaction, in the accuracy of the models and the methods we developed to interpret and apply them. I hate the sick sense of recognition of reading newspaper headlines reporting impacts predicted decades ago in scientific papers. I experience a sense of destabilizing déjà vu when sce-narios I was assigned as an exercise in graduate school are hap-pening in real life. It's horrible that nightmares have gone from prediction to reality, and it's terrifying to contemplate what the models say about the 3°C-plus world we're heading for under cur-rent policies.

Looking back on my career so far, I can identify so many ex-amples where I haven't raised my voice as much or as plainly as I could have, where I haven't sounded the alarm loudly enough. I don't intend to keep making that mistake.

May Cooler Heads Prevail

Okay. Deep breaths. My feelings of outrage and discouragement notwithstanding, there is some good news here, and it comes from the data collected by subject experts, whom I trust to inform me about the world.

One bright spot is that people who deny climate change are awfully loud, but they're not actually that common. In the United States (where much climate change denialism originates), a November 2019 Yale survey classified 10 percent as "climate dismissives," who do not "believe" the fact that humans are warming the climate.

What drives climate change denialism? Research led by my psychologist colleague Kirsti Jylhä shows that the strongest predictors are beliefs that justify and promote existing hierarchies: the human domination of nature, and exclusionary, anti-egalitarian social views (for example, opposition to feminism and multiculturalism). This shows that the Exploitation Mindset placing man over nature, and some humans over others, is indeed a root cause of our current mess, and overcoming this mindset with one based on empathy and respect will be critical to moving forward.

Katharine Hayhoe, an evangelical Christian climate scientist, says that we can and should have constructive conversations with the 90 percent of people whose personal identity is not "built on a set of beliefs that reject the reality of changing climate." She advocates starting with shared values and interests (like faith, or birdwatching), not facts, to establish common ground. A 2020 study on understanding and overcoming climate change denialism agreed, advocating discussing values first, focusing on scientific consensus, emphasizing the purity of the Earth rather than how humans treat it, and reframing solutions as upholding social longevity (rather than threatening change).

A 2020 *Nature* study reported that a survey in Oklahoma found that climate beliefs of partisans on the political right, though much less supportive of climate action, were also much less stable than those on the left. The authors concluded that political beliefs about climate change "maintain the potential to shift toward broader acceptance and a perceived need for action." In another study, communicating scientific consensus (the statement "97 percent of climate scientists have concluded that human-caused global warming is happening") depolarized respondent judgments about climate change. Facts do matter.

Happily, according to Pew Research Center surveys, the American public overwhelmingly does trust scientists. Further, that trust increased 10 percent in the latest three years, with 86 percent of those surveyed in 2019 expressing "a great deal" or "a fair amount" of confidence in scientists to act in the public interest. (Aw, thanks, peeps, we love you too! Even though we haven't forgotten how you made fun of what giant nerds we were when we won that sixth-grade math contest.) About twice as many Americans had confidence in scientists compared with the media or business leaders. Elected officials, whose core duty is to act in the best interest of the public, were the group that inspired the least public confidence to do so, at just 35 percent.

More good news for science: In recent years there has been a surge of increased public understanding in line with the science. The Yale "Climate Change in the American Mind" surveys show a sharp increase since 2013 in the percentage of Americans who know global warming is happening, know it is caused by humans, and know that scientists agree it's happening. In 2018, 73 percent of Americans thought climate change was happening and 72 percent said it was important to them personally—a powerful majority. Nearly as many (69 percent) are worried about it and agree it will harm Americans (65 percent). Just short of a majority agreed "climate change will harm me personally" (49 percent) or "I have already experienced its effects" (46 percent). (For that last group, I have some homework: Read the 2018 *Fourth National Climate Assessment* by federal government agency scientists, showing "the impacts of climate change are already being felt in communities across the country.")

Despite decades of climate delay and misinformation, far too slowly but perhaps just in time, the truth is winning out. Now we have to act on it as if our lives depend on it.

Harnessing Feelings for Action

Anger at those who practice deliberate harm and deception is justified. Anger toward dishonest leaders and lobbyists is righteous. My friend Ben says that anger signals when our principles have been violated. Being angry at injustice is natural. (If you want to see a great example of righteous anger, look at the Norwegian clip of the faces of the girls who do the same work as boys in a sorting task but receive half the candy reward. To the boys' credit, they recognize the injustice and divide the candy evenly.)

Personally, I find anger can be empowering, if I use its rush of energy to fuel organizing and action, which is what we're ready to move on to for the rest of the book. I'm going to start with some of the systemic climate injustices that make me the angriest—but I'll also do some uncomfortable looking in the mirror to examine my own accountability.

In Part 2 we've been on an emotional roller coaster through the huge range of feelings that we face in the climate crisis. We've examined how we can use the crises of our atmosphere and earth as a lens to clarify what we value most and to set us on course toward purpose and meaning by putting those values into action. Identifying and directing your talents toward the meaningful work of protecting people and nature, reducing harm, and increasing resilience throughout your spheres of influence is how to build regeneration and dismantle exploitation. Acknowledging and honoring All the Climate Feels strengthens our empathy, collaboration, and resilience. I hope it will help you renew and sustain your energy for the marathon ahead, and enhance the joy that comes from nurturing and fully experiencing relationships with nature, one another, and ourselves. But now it's time to use these feelings to power action.

Part III

We Can Fix It.

How We'll Go Forward

Chapter 8

Climate Change Isn't Fair

Acknowledging Climate Privilege

At the Copenhagen chapter of the global People's Climate March in September 2014, I watched an eight-year-old blond girl bravely take the microphone in front of a large crowd. Surrounded by tiny friends wearing striped leggings and carrying signs and banners painted with green hearts, she said, "I dream of studying the ocean, but I'm afraid the ocean may be dead when I grow up."

The following year, as an observer to the 2015 Paris Agreement negotiations, I remember hearing one of a handful of delegates from a small island state ask the chair to structure the negotiations differently. As the deadline for reaching an agreement approached, parallel negotiations for different parts of the text were being hammered out in half a dozen rooms simultaneously. This was no problem for the suited army of more than a hundred representatives from the United States, but there was no way the members of the tiny delegation could be everywhere at once. This meant they were missing out on giving input to decisions that could literally affect whether their homeland will survive sea level rise or not. The residents of small island states have collectively emitted nearly nothing, but they are poised to lose everything from climate change.

Who needs to do what to stabilize the climate? There are lots of ways to try to understand and allocate climate responsibility. You can slice the carbon pie by country or by industry, looking at the past or the present, production or consumption. All these emitters have important responsibilities. But no matter how you look at it, I believe what high emitters in rich countries, especially Americans, do individually and collectively in the next decade is absolutely critical to the climate of the whole world, forever. This means that individuals have much more power to make meaningful change than most of us currently realize. We need to start seizing this power and putting it to use.

The Cruelty of Carbon

Climate change is heartbreakingly cruel. Carbon emissions have been obscenely unevenly distributed across and within nations, by income, and over time across generations. The disparities between those who have caused the most emissions and those who suffer the greatest impacts deeply entwine climate change with broader social justice issues, from economic, gender, and racial equality to public health and education.

Genuine solutions to all these issues need to address root causes. Policies to reduce gender and income inequalities, as well as those to address poverty, were recently shown to most benefit other sustainability goals like zero hunger, biodiversity, and climate action.

Piling climate impacts on an already unfair world both shines a light on and worsens existing injustices. These impacts tilt the playing field further against those scrambling to overcome systemic barriers and disadvantages, and make it harder for people to achieve their full potential, thereby prolonging human suffering. Worsening climate impacts also divert resources to respond to and recover from increasingly frequent disasters, meaning those same resources are no longer available for planning ahead to reduce their likelihood in the first place. It's hard to act with foresight to reduce emissions along with non-climate risks when you're racing around trying to put out multiple fires at once.

An example of how climate change exacerbates inequality comes from my friend Sara Gabrielsson's work to get reusable menstrual cups to teenage girls in rural Tanzania. Studies show that empowering and educating women and girls is a key strategy to support human development. (As a side benefit, it also tends to reduce emissions of greenhouse gases.) In many poor countries, though, it's common for teenage girls to drop out of school because they can't afford menstrual hygiene products. As the short film *Breaking the Silence!* shows, desperate girls try using corncobs, leaves, or bits of mattress stuffing as a makeshift pad. At school, there is no access to toilets or running water. At home, the girls are responsible for fetching water for their families to drink, cook with, and bathe in, a chore that takes hours each day. Climate change is drying out these water sources, demanding ever more time to walk farther for water collection, and making these regions poorer, farming life harder, and the chance of education more remote.

Of course, in an age of shocking and widening income inequality, we don't have to go overseas for infuriating examples of disparity; there are plenty within the United States. Hurricane Katrina struck New Orleans in 2005. Eight years later, the median Black household income in New Orleans was $5,000 *less* than it had been in 2000 (adjusted for inflation), widening the gap between Black and white household earnings. White neighborhoods tended to be built on higher ground and suffer less damage, while Black households often lacked the wealth to rebound and faced discriminatory rebuilding programs. The result was a permanent exodus of nearly 100,000 Black residents.

Some Countries Are More Equal than Others

Despite inequalities and differences, the world stands behind the Paris Agreement, which says that solving climate change is in the shared collective interest of humankind, and everyone must do all they can, given their circumstances, to help make that happen. Each country is responsible for making ambitious contributions to help meet the collective goals. All countries must reduce emissions, but their contribution to the problem and their capacities to address it, and therefore their obligations, vary.

The Paris Agreement is a framework for coordinated global climate action, built on a process where countries make pledges to reduce emissions, including specific targets for how much to reduce emissions by when, then implement policies to achieve these reductions. The country pledges are continuously reviewed and critiqued and are updated every five years. For countries to meet their pledges, more and better government policies are needed. More ambitious policies rely on and are supported by civil society oversight and pressure as well as institutional leadership, as we'll come back to in Chapter 12.

That's the idea, anyway. So far, "Governments [are] still showing little sign of acting on [the] climate crisis," as the title of Climate Action Tracker's December 2019 report pithily summarized. Their analysis of national climate policies covering 80 percent of global emissions found that just one country (the Gambia) has both a Paris pledge compatible with limiting warming to 1.5°C and policies in place to meet this goal in practice. (Morocco's policies are judged "close" to meeting 1.5°C.) A further handful of countries have pledges consistent with 2°C, though again, only two have policies judged sufficient to deliver these pledges (Kenya and India, the world's fourth-largest emitter). The rest of the national pledges, including all of the remaining major emitters, are judged "insufficient" (the European Union), "highly insufficient" (China), or "critically insufficient" (Russia, Saudi Arabia, and, last but not least, the good old US of A). So right now, a few countries that are largely causing very little climate pollution are most ready to act, while the countries with the most historical responsibility and the most current climate pollution to reduce are shirking their duty. It's not a good look.

Following the Regeneration Mindset principle that all people have equal value, and recognizing that the carbon budget is a collective treasure and legacy of the entire human family, the fairest way to divide the carbon pie is that each person, over time, has an equal right to the same share.

But a fundamental inequity at the heart of our global climate problem today is that people in the United States and Europe have historically and continue still to vastly overconsume their fair share of the carbon budget; together, they have emitted nearly half of the cumulative carbon now warming our atmosphere. On an annual basis, Europe, starting with the United Kingdom, emitted 90 percent of fossil fuel emissions in 1850; the United States took over as the world's leading annual emitter not long after, only surpassed by China in 2006, which through 2018 had contributed 13 percent of the cumulative total. Western leaders who throw up their hands and proclaim, "It doesn't matter what we do, look at China!" remind me of someone eating nearly all the cake, then blaming its disappearance on the person reaching for their first piece.

If we look at per capita emissions, the story is stark: A very few have polluted the most. If rights to the carbon budget were distributed equally, emissions would match population. That is, if residents of a country represented 5 percent of the global population, their emissions should also represent 5 percent of the global total. This is not remotely the case. The overconsumption of the United States is especially shameful, where about 5 percent of the human family has managed to emit 25 percent of its carbon to date. Meanwhile, the residents of all African countries represent about 15 percent of the world's people but have emitted only 4 percent of the cumulative carbon. India divides the same 4 percent of cumulative emissions over a larger population, making their per capita emissions ten times lower than those of the United States. China's per capita emissions are less than half those of the United States.

The vast carbon discrepancies between rich and poor countries illustrate a more general trend: A study led by Daniel O'Neill showed there is *currently no country on earth* that meets human needs within ecologically sustainable resource consumption limits. Basically, researchers found that rich countries mostly manage to meet human needs like nutrition, sanitation, healthy life expectancy, access to energy, and education, as well as employment, equality, and life satisfaction. However, rich countries vastly overconsume our share of finite natural resources, including land, water, and materials. Meanwhile, despite substantial advances in human progress in the last few generations, poor countries generally lie within the safe operating space ecologically but still fall critically short on social goals. The authors found that 82 percent of the countries studied lacked sufficient democratic quality, two-thirds lacked access to clean water and toilets, and more than 40 percent had insufficient food.

If you are lucky enough to be one of the few with your human needs largely met, congratulations! But here's an uncomfortable truth: One reason the people Over There are suffering more than you are right now is because *your lifestyle* takes up such a disproportionate share of the carbon budget and other finite resources. People in countries like the United States and Sweden are using resources like there are four planet Earths, leaving less than nothing for everyone else.

From Countries to Households

Seventy-two percent of global climate pollution can be traced to household consumption decisions, including mobility (especially using cars and planes), diet (especially meat and dairy consumption), and housing (heating and cooling and electricity consumption). The remaining emissions come from government spending, infrastructure, or capital investments (where we as individuals can also play a powerful role in system change, as I'll discuss in Chapters 11 and 12).

"Wait a minute," I hear you say. Didn't I just say in the last chapter that a small group of fossil fuel companies are responsible for more than 70 percent of industrial emissions? And now I'm saying households (mostly wealthy ones) cause more than 70 percent of the total? That adds up to more than 100 percent!

You are absolutely right, astute reader! Both of these statistics are correct, depending on whether you allocate responsibility to the producer (fossil fuel companies) or the consumer (households, governments, and investments). In either case, it's the same fossil fuels being dug up, and the same land-use change being driven. We are part of the same system, and the whole system needs to redirect from exploitation to regeneration. Both production and consumption of fossil fuels have to end.

Today, income buys climate pollution. I and many of you reading this—for example, anyone in the United States making over $38,000 a year—have incomes among the top 10 percent in the world. Collectively, we are responsible for half of household carbon pollution worldwide. Meanwhile, the poorest 50 percent of people in the world collectively cause just 10 percent of household carbon emissions. Most of their tiny emissions are for necessary survival activities like cooking and home heating.

The disproportionate allocation of climate responsibility continues within the United States, where an Oxfam report showed that individuals in the poorest 50 percent of households (think janitors, grocery clerks, and other essential workers) cause about eight tons of carbon pollution per year. These workers have limited ability to cut emissions, both practically (their emissions are already low!) and financially (they are often living paycheck to paycheck). In the median household (living on about $69,000 per year in 2019, roughly the income of two preschool teachers or one nurse), per capita emissions are about sixteen tons per year.

The richest 10 percent of Americans (with household income levels above about $201,000 per year), a group that includes many of my friends from college working in management and professions like law, medicine, and tech, emit an average of fifty (fifty!) tons per person per year. While emissions from food stay relatively constant, travel by car and especially by plane makes up an increasingly large share of personal emissions as income rises. The super-rich top 1 percent are even more extreme; my Lund University colleague Stefan Gössling estimates that Bill Gates and Paris Hilton emit an astonishing ten thousand times more carbon from flying than the average person.

The Rich Need to Get to Work

Meanwhile, the budget to limit warming to 1.5°C is 2.5 tons of climate pollution per person per year for household emissions by 2030, according to an Aalto University report. Keep that number in mind for where we all need to be aiming. A 2020 study led by Diana Ivanova showed that only 5 percent of European households live within that budget today. If we imagine the recommended daily calorie limit as a comparison to the personal budget for 1.5°C, Americans' current carbon overconsumption is like the average American eating thirteen Big Mac Meals per day, and wealthy Americans eating forty Big Mac Meals per day. Gross.

The good news is the changes needed are achievable. The Finnish future fund Sitra has profiled how a combination of political decisions, company offerings, and personal choices could lead to people living well within 2.5 tons per year by 2030. Far from the accusation that we'll all have to go back to cave dwelling, a 2020 study found the world could use 60 percent less energy than today and provide a good standard of living for everyone: comfortable housing with clean water, a washing machine, fridge/freezer, laptops, and smartphones; hospitals and schools; and transit to provide 5,000–15,000 kilometers of mobility a year. But to make the energy enough to go round, today's excessive consumption by the affluent would have to be reined in.

With their emissions already so low, large personal footprint reductions are impossible for the very poor. They are unnecessary for the global middle class (residents of India are currently below the universal per capita 2030 budget of 2.5 tons; Brazil is at 2.8 tons).

But reductions in luxury personal emissions from high emitters like me? Those are imperative.

Professor of energy and climate change Kevin Anderson has calculated that, if the richest 10 percent of the global population reduced our emissions to the level of the average European (a comfortable standard of living), we would cut *global* emissions by one-third. That's most of the way to the 50 percent cut needed by 2030. While building infrastructure and passing and implementing policies take time, the global rich can cut our unnecessary emissions fast. Remember that, to actually decrease emissions, we have to shut down fossil energy and replace it with clean energy, whereas historically, we've added clean energy on top of dirty to meet increasing demand. Reducing demand from overconsumption is critical to a fast clean-energy transition.

Here's the point: The imperative to fairly share the carbon pie means that those of us lucky enough to have drawn "high-emitting country" as our place of birth in the cosmic lottery, and who have benefited the most from historical fossil-fueled development, have a responsibility to reduce our emissions first and fastest. Importantly, this means that while the world needs to cut *global* carbon emissions in half by 2030, *emissions in places like the United States and Europe should be cut by more than half by 2030,* because poorer countries with lower emissions have both less emissions to cut in the first place and less ability to do so.

A key element of climate justice is for high-emitting countries and individuals to rapidly reduce our own emissions. By doing so, we leave more space for people who need their emissions to survive, *and* we lessen their burden in facing the impacts of climate change they haven't caused. (High emitters reducing emissions fast is part of our "common but differentiated responsibilities," in UN-speak.) The possibility for rich countries to transfer carbon-cutting tech to poor countries is often cited as a way to cost-effectively reduce emissions while helping the most vulnerable address climate change, but sometimes the essential contribution of high emitters ourselves cutting emissions is overlooked. The Caribbean was struck by record-breaking Hurricane Irma in 2017, which scientists said was consistent with human-warmed waters driving stronger hurricanes. As his country struggled to rebuild from the devastation, Gaston Browne, the prime minister of Antigua and Barbuda, was clear about what his citizens ultimately needed most: "In fact, the best way that large countries who rely on fossil fuels can assist us is to reduce emissions."

The good/bad news: If you live in a rich country, and especially the richer you are, your individual actions really matter. They are important materially to the amount of carbon in the atmosphere, because they tackle one of the major sources of climate pollution globally. Your behavior also helps change the culture, showing that the vision of a good life is transforming from a high-carbon life to the low-carbon high life. Further, practicing what you preach enhances your moral authority and ability to inspire others, as I'll talk about shortly.

Personal Carbon Responsibility

As I showed in the previous chapter, there is abundant evidence of the infuriating, devastating delay and deception of the fossil fuel industry.

But! I also have to face my own carbon responsibility, and there are a lot of people like me who have been using way more than our share of the carbon budget—many more of us than there are fossil fuel execs. Cringe. One major reason the world is decades behind where we should be on climate action today is that well-meaning, altruistic, middle-class-plus people like me haven't meaningfully changed the signals we send in the marketplace with our overconsumption of high-carbon luxuries. As a result of our inaction, powerful companies and politicians have successfully argued that defending our overconsumption is more important than making the necessary changes to stabilize the climate.

If we're going to solve climate change, the carbon elite have a critical role in leading the change. So how do we do that? First, I would argue, by acknowledging our own climate privilege.

When I first came to terms with the fact that I had been using up far more than my fair share of the carbon budget, it made me realize that I have a greater responsibility to act decisively to reduce my own emissions. One way that I can lead change is by being the change I want to see, starting with where I am and what I can control.

For those of us at the top of the income ladder, our current overconsumption (of fossil fuels, of unhealthy and unsustainable food, of stuff) is actually making us less happy and healthy, and creating anxieties of its own. We're hurting ourselves as well as the planet. The rich world needs to transform toward something worth emulating and passing on, not setting our unsustainable wastefulness of the past as the aspirations for what a good life looks like. We need to figure out how to be not just consumers, but citizens, and see how these changes are part of creating a cultural shift centered on core values we are proud to live out and pass along.

For rich people like me, this means we need to consume and pollute radically less, and simply be radically more. Focusing on getting your own life in carbon order is a win-win for all: your children, other people's children all over the world, and finally—surprise—you, who might discover that your new, sustainable life is lovely and rich in a new way.

The carbon elite need to change, and I believe people are willing to change. I don't believe the vast majority of people value short-term comfort over longer-term survivability. People make changes and even sacrifices all the time for people they love and for causes they care about and believe in.

Climate doomers believe that it is human nature for people to turn money into high-carbon luxuries of convenience that they are unwilling to give up, even if the price is a habitable planet. Roy Scranton speculates, "The freedom to pollute is a kind of status symbol. . . . We can assume that the wealthy parasites killing the planet are never going to give up their privileged and destructive habits, because those very habits are how they maintain their sense of self-worth."

Given the correlation with income and pollution, Scranton is writing this description about me, my friends, and almost certainly, as an English professor at a major university, himself. I don't believe his view of human nature for one minute. When faced with the trade-off between taking ten flights per year and finding some lovely holiday spots closer to home and helping to prevent the extinction of all coral reefs on Earth (which would take millions of years to re-evolve, if ever), will actual human beings really choose their frequent-flyer miles? When faced with the option to add more healthy years to your life and prevent contributing to the sixth global extinction, or keep eating the food that is killing you and the planet, does the second option really sound better? Are we truly so addicted to buying and throwing away crap that we're willing to condemn the whole planet and all of humanity from now forward to an impoverished and dangerous existence? Wouldn't it be better, as Juliet Schor says, to have more fun and less stuff?

Both/And, Not Either/Or

The research is clear: Both individual and collective actions, in both private and professional life, are needed to reduce emissions toward zero fast. Fortunately, personal and collective action are deeply entwined with and reinforce each other.

But talking about the need for personal carbon responsibility earns pushback. Some climate justice advocates argue that placing climate responsibility on individuals unfairly focuses the burden of change on already marginalized communities, while letting politicians and businesses off the hook. It's true that some businesses embrace personal footprints to offload their own responsibility. And that politicians afraid of losing voters have avoided treading on "lifestyle issues." As George H. W. Bush famously declared at the UN Earth Summit in 1992, "The American way of life is not up for negotiation."

Don't worry, I'm not letting the titans of industry and politics off the hook. Fundamental, system-level economic and political changes are needed to tackle climate change; they have their own chapters coming right up! But personal changes *from high emitters* (not marginalized low emitters) are also needed; it's both/and, not either/or.

The argument that the only changes needed are political and not individual is one way of denying responsibility. I fully agree that those with low incomes are constrained in their ability to make lifestyle changes—but they don't need to; these are the individuals who already produce the lowest emissions. I'm less sympathetic to relatively well-off carbon overconsumers arguing they'll change if and only if everyone else is forced to change too. For example, after citing the climate damage of flights and burgers, the climate journalist Amy Harder wrote that she usually flies more than six times a year and eats beef a couple times per week, but "I'm not losing sleep over my flying and eating habits—and I'll only make big changes if the price tags get a lot bigger." That's another way of saying, "I'll only do the right thing when I know I'd be the only one continuing to do wrong."

We will not solve climate change without reducing overconsumption from high emitters. There is just no getting around the fact that quickly zeroing out emissions will require carbon overconsumers to cut their pollution down to a healthier level. Policies can and should support these transitions, but there is an enormous role for those who can to lead by example, as a part of driving cultural change.

Don't Sweat the Small Stuff

Knowing that individual climate actions are critical, especially for high emitters, motivated a 2017 study I published with my colleague Seth Wynes. What we found in "The Climate Mitigation Gap" is that, for high emitters, the big actions that can cut today's emissions fast are going flight-, car-, and meat-free. These are the personal actions that contribute most to the urgent planetwide need to get to the root of our problems and shut down the systems of fossil fuels and industrial animal agriculture like we mean it. Focusing on cutting your emissions in these three domains is where your personal carbon footprint effort should go. Start with the one that feels most feasible for you; if going totally without isn't possible, aim to cut it at least in half. These actions are the personal versions of the systemic changes needed to overhaul the transport and food systems to stabilize the climate.

If you want to dial into your personal impact even more, you can take a carbon footprint quiz. The CoolClimate Calculator from University of California, Berkeley researchers is the most evidence-based one in the United States. If you want a pretty design and more user-friendly focus on high-impact actions, try the lifestyle test from Sitra (it's made for Finland, but general principles apply). These quizzes can be eye-opening and are a good way to establish a baseline before you dive into making changes. But if you want to skip the calculators, based on our study compiling thirty-nine studies and carbon calculators, I promise if you start with reducing flights, then driving, then meat, you'll be having the biggest bang for your climate buck.

A major benefit of focusing on high-impact actions is that you can concentrate your precious attention on what really makes a big difference. This is especially important because research shows most people have no idea which of their lifestyle choices in fact have the biggest climate impact and thus often end up investing time and energy trying to change behaviors that don't substantially cut carbon. Even the smallest of the high-impact climate actions we identified (eating a plant-based diet) makes a far bigger difference for the climate than more common but much lower-impact choices like recycling everything in your household or using a reusable shopping bag for a year. In fact, we found a plant-based diet saves four times more emissions than recycling, and 160 times more than a reusable bag. If you've already made these small daily actions part of your routine, great! Keep going, but you're ready to level up to tackling high-impact climate actions.

A lot of well-intentioned campaigns use the foot-in-the-door, every-little-bit-helps approach, reasoning that if everyone undertook a small action like avoiding plastic straws, it would add up to a big impact. To me, this thinking is exactly backward. An action with climate benefits so small that it requires everyone to participate to make an impact is where personal behavior change is *least* effective (instead, defaults and standards are needed). In terms of creating ripple effects, research shows that high-commitment actions are more likely to spill over to additional actions. Unfortunately, there's little evidence to suggest that starting with one tiny thing, like recycling, actually motivates people to take on the next, necessary, bigger steps. Good intentions demonstrated through small actions may even backfire, as people feel they're "already doing their share" by recycling, and thus rationalize continuing their multiple flights a year and long drives, even though the carbon math doesn't remotely add up.

While the purpose is always to reduce harm at the source and increase resilience while protecting people and nature, one key insight from our study is that the most effective actions for the climate depend on who is undertaking them. Effective policy needs to take the responsible actor into account. For example, reducing food waste is a leading climate solution at the global scale, because the issue is widespread but diffuse: Everyone needs to eat. Globally more than one-third of food goes to waste; in poor countries, primarily from lack of harvest and storage, and in rich countries primarily by consumers, where waste prevention through better planning is key. For issues like this, where widespread behavior and systemic changes are needed at scale, regulation, standards, and nudges will be more effective. No one likes wasting resources and money, so individual efforts to reduce food waste are laudable, but our research shows eating a plant-based diet saves twice as much climate pollution as a year of eliminating personal food waste. Reducing overconsumption of high-emissions products and services beats improved efficiency of using them, because it gets to the root cause of the issue.

What Makes Change Happen?

My theory of change is grounded in academic research, but my personal experience of seeing what makes change happen is my most powerful motivator. I can report that it was personally transformative to align my behavior with my values. As I'll share in the coming chapters, I was inspired most by seeing the actions of people around me who led the way with their own behavior. (In hindsight this makes sense; see lifetime success of "Do as I say, not as I do.") Research shows this is how social and cultural change actually happens and scales up, from person to person until it reaches a critical mass where the whole culture is affected. Starting with myself, with what I could take responsibility for, also expanded how I saw myself. It empowered me to see the role I played in much bigger systems and to start using the power I already had to change them.

One more reason to support personal action is that excessive reliance on tech, or complacent belief in techno-salvation, tends to outsource responsibility to engineers or other distant experts, which can give false hope and thereby prevent meaningful action. A study headed by Jennifer Marlon showed that the unrealistic expectation that an external force (like technology, Mother Nature, or a higher power) will solve climate change decreased climate policy support and political engagement. The researchers also found that believing it's impossible or too late to fix the climate crisis led to "fatalistic doubts," which paralyze action too.

One of the biggest barriers to getting the climate action ball rolling is that *everyone* is waiting for someone else to go first, while simultaneously feeling like their own actions don't matter. Obviously, these two things cannot be true at the same time.

Surveys show that large majorities of people care about the climate and are willing to act to help fix it. One recent survey found 80 percent in the United States and United Kingdom are willing to make lifestyle changes for the climate as big as the ones they've made for coronavirus. Young people especially value a stable climate over conveniences and luxuries; a recent Swedish survey showed that 85 percent (!) of youth aged twelve to eighteen were willing to make lifestyle changes, particularly to travel less by car and plane. But right now most people are paralyzed, believing they're alone—when in fact we're a powerful majority.

Research shows that two things motivate people to take meaningful action. The first is internalizing knowledge of the seriousness of the climate crisis and accepting their share of personal responsibility. If it appears to be someone else's problem, people don't take action. Basically, people care about doing the right thing and feel better when they do.

The second big kick in the pants is seeing others actually act. As humans, we follow and are inspired by others' behavior. Climate actions like avoiding flying or getting solar panels are socially contagious and help people feel empowered to make necessary changes. Behaviors also ripple outward to shape culture.

Personal behavior change can be really powerful. We want to know that our actions have consequences. And they do. To "be the change you wish to see in the world," to misquote Gandhi, not only gives moral strength and increases personal meaning and motivation, but also strengthens community and amplifies positive change. Putting your values into action in your behavior can reduce anxiety and cognitive dissonance, and it can also inspire others to act.

I love the way that my colleague Kim Cobb described biking to work and, later, cutting her flying 90 percent as part of her work for climate justice, which has certainly inspired many others: "Having done the math on my carbon footprint, that was just not a number I could un-see. . . . I started biking to work, and this was one of the sparks that helped me understand the deep rewards that can come from what we might call individual action. . . . When I was locked in my car for fourteen years in Atlanta, I could not have foreseen this package of personal rewards and these ripple effects. It all looked like sacrifice to me. . . . [But] these things are possible. I like to think about what our 2050 world will look like, and enact that today."

Responsibility scales with power. Certainly, there are people with much more power and responsibility than I have: deceptive oil executives; politicians who have kicked the climate can down the road for election cycle after election cycle. But that doesn't let me off the hook for being accountable for doing what I can to help prevent climate breakdown now, starting with where I am.

Chapter 9

Slowing Down and Staying Grounded

Embracing a Car-Free, Flying-Less Life

C limate confession time: I am a recovering frequent flyer. In my twenties, I dreamed of traveling to exotic destinations (or perhaps more honestly, of being someone who had been to such places). If my friends or I ever had any free time or money, our instinct was to turn it into a far-flung, high-carbon holiday, like hiking the Inca Trail on spring break. We would share tips about how to best hack air miles to turn them into more trips. When social media came along, we used it to post evidence of our well-traveled world citizenship. We were all about Experiences over Things.

Voluntary simplicity and minimalism have big upsides. Westerners are drowning in our things and in the stuff we throw away. But while our consumption of goods does have an impact, the climate pollution from our less visible consumption of transport and electricity is actually much worse. Plastics (which are made from fossil fuels) pollute ecosystems and harm wildlife, and are a symbol of unsustainable production and overconsumption. The plastics sector needs to become fossil-free and zero emissions along with everyone else. But even studiously avoiding plastics saves way less emissions than the high-impact personal actions of cutting cars, planes, and meat.

I was winning at Experiences over Things as I dutifully schlepped my reusable water bottle on lots and lots of flights, but the climate was losing: One flight canceled out the climate savings from ten thousand avoided plastic water bottles. In 2010, my peak flying year, I took fifteen round-trip flights, for a mix of work and pleasure: interviewing for that increasingly rare unicorn, a tenure-track faculty job; meeting colleagues at workshops or conferences; a stunningly cheap weekend trip to Disneyland with friends.

After three years on the academic job market, and applying to sixty-six faculty jobs, I was offered one of the coveted tenure-track positions. The catch: It was in Sweden, an eleven-hour flight away from my family and closest friends in California.

At that time, I didn't consider the carbon implications of placing myself on the other side of an ocean from my nearest and dearest (nor did I entertain the possibility that a global pandemic could ground aviation to an abrupt, if temporary, near halt a few years later). I just reassured myself I would come home often on one of many frequent, cheap flights, and I took the leap.

Just a few weeks after I moved halfway across the world to Sweden, I found myself unexpectedly single for the first time since *Pulp Fiction* was in theaters. My immediate instinct after ending a sixteen-year relationship was to buy a plane ticket to escape. I booked a solo divorce-moon to nurse my broken heart, and umbrella drinks before noon, at an Egyptian beach resort on the Red Sea. But I was starting to realize that I couldn't outrun my worries by hopping on a flight to some exotic new locale.

A Fateful Beer in Vienna

My high-flying days continued for a couple of years after I moved to Sweden, but I was becoming increasingly uncomfortable with the yawning gap between the urgency of the climate emergency and my own high carbon footprint.

It took a long talk with my friend Charlie and some liquid courage to confront my cognitive dissonance around my frequent flying. We were attending a climate conference in Vienna in 2012, where we'd spent ten hours in a windowless room being assaulted by yet more data showing how bad climate change is and how urgently we need to reduce emissions.

But the hypocrisy that I, and most of my fellow attendees, had flown there did not escape me. Hour for hour, flying is the fastest way to heat the planet. I was starting to feel like part of the problem. A mounting pile of studies demonstrates that scientists who lead low-carbon lifestyles themselves are more effective in inspiring the public both to change their own behavior and to support ambitious climate policy.

One of the major job perks of being an academic, and a not insignificant part of my initial motivation to become one, was the chance to go to conferences in new and interesting locations, to have nice meals and drinks with familiar faces against an exotic background. The idea that researchers must improve our reputation by presenting at conferences is baked into the culture of academia; international presentations are criteria for job promotion. As a grad student, I admired the successful professors who claimed to no longer have a circadian rhythm for jet lag to mess up because they were so used to switching time zones every week. Someone who commuted weekly between his Harvard professorship and the consultancy he ran in Mumbai was spoken of exaltingly. Right after your list of publications and grants, your gold-card lounge status was a proxy for your intellectual contribution and importance.

But in Vienna, I was starting to feel like I was attending a conference of doctors puffing on cigarettes while telling our patients to quit smoking.

Charlie was different. He'd come to the conference by a fifteen-hour train trip from the United Kingdom, which he described with enthusiasm over a cold beer.

"I can't imagine never flying to see my family again. How do you make your life work without flying?" I asked him.

"Look, I haven't given up flying completely. But I decided, if I can go where I need to without flying, I do. I've made this one big decision that saves me tons of effort from making lots of tedious, tricky decisions about how to travel from A to B. Meanwhile, I've intentionally wound down my work commitments in North America and increased my collaborations in Europe, to reduce my need to fly. It's like deciding to be vegetarian: When you go out for dinner, you're spared the continuous thought of 'Ooh, should I have the steak or not?' But if I was starving in Argentina and the only food available was steak, I'd eat steak. Not that I've ever been to Argentina!"

I felt a light bulb go off in my brain. I could do that too.

Until that moment, I never imagined all the flying I *could* reduce, because I would get stuck on the one flight I couldn't bear to imagine giving up: the one from Sweden to California to see my family, one of the journeys George Monbiot calls "love miles." But now a middle ground appeared, something I could commit to. I couldn't do everything right away, but I could start with something.

After this talk with Charlie in 2012, I became someone who did not fly within my continent. This forced me to think about myself and my time differently. Knowing it would take longer to get someplace made me reconsider the value of my time: Did this thing really need to be done? If so, am I really the best one to do it? Considering these questions has made me turn down a lot more travel, or switch over to digital presentations. For example, when I was recently invited to give a keynote lecture in Spain, my hosts were happy to pay for me to professionally record my lecture, which I've now shared online for anyone to watch, at a much lower carbon and financial cost than me physically traveling between countries for a one-hour occasion. I only agree to trips I feel are really important, where I could make a unique contribution. In these cases, I find a way there overland, usually by train, often planning stops to see friends and collaborators along the way to make the best use of the journey.

Making the decision not to fly within Europe was enormously relieving. In trying to align my actions with my values, I hadn't realized how much energy I was wasting agonizing over daily, tangible but trivial distractions that weren't a big deal for my carbon footprint, like shrink-wrapped cucumbers in the grocery store. Meanwhile, I had been performing exhausting mental gymnastics to try to justify why saving a few hours of my time was worth months of my sustainable carbon budget, when at heart I just couldn't. It was liberating to have made a few key decisions about things that really mattered, to the climate and to me. I'd been so stuck in my status quo I hadn't seen its downsides: the stress and exhaustion of the feeling of constantly trying to keep up that led me to try to be everywhere and nowhere simultaneously. As a former frequent flyer, now down to at most one flight a year to see family in North America, I have cut my flying emissions about 90 percent.

Today, I don't miss dragging myself bleary-eyed to the airport and home late that night to make a one-hour meeting a thousand miles away. I've still been a productive researcher, and I've still gotten to explore new places. My new approach has led to unexpected discoveries while traveling, like surprising adventures in tiny museums waiting for train connections, as well as bonus visits with friends, and even romance (more on that in a minute).

Staying on the Ground

Globally, emissions from flying can look small, but that's because it's such an elite activity: only 11 percent of the world population flew in 2018. More than half of adults in the US and UK, and more than two-thirds in Germany, never fly. Meanwhile, the 1 percent of the global population who are frequent flyers cause *half* of air travel emissions. In the US, 68 percent of all flights are taken by the 12 percent of adults who take six or more flights per year.

When I was a frequent flyer in 2010, my flights alone were about twenty-five tons of CO_2. That is ten years' worth of my sustainable carbon budget. Put another way, my flying that year alone used up the carbon budget for ten people to meet all their needs. Squirm.

Here's a fun, true fact: You use up more of what should be your lifetime carbon-expenditure budget while mindlessly watching four rom-coms on a plane than you save in *four years* of carefully and accurately recycling *everything you throw out*. It's true. I shit you not. I am sorry.

Our 2017 study found that a single round-trip transatlantic flight, like New York–London, emits almost eight months of your sustainable carbon budget—those two and a half tons per person per year the whole world needs to be at by 2030 to be on track to limit warming to 1.5°C. To equal the emissions from that one return plane trip, you'd have to skip meat for two *years,* or eight months of typical driving (though just six months of typical American driving, where bigger, less efficient cars and more miles traveled make the driving footprint more than twice as high as in the United Kingdom).

Similarly, while emissions from tourism cause about 8 percent of global climate pollution, for wealthy individuals, aviation alone can contribute 50 percent of their household carbon footprint. Not unusually for frequent-flying professionals, the climate scientist Peter Kalmus found more than two-thirds of his emissions came from flying, which he stopped doing in 2013. From this and other lifestyle changes, he now lives on one-tenth the fossil fuel of the average American. All emissions need to head toward zero, so every source of emissions needs a plan to do so, whether that's Germany (2 percent of global emissions) or the aviation industry, which emits 2.4 percent of global carbon and also has high-altitude effects that make their warming equivalent to 7.2 percent of global carbon emissions. At the personal level, for high-income households, airplane travel dominates the climate pollution picture.

In 2018, the aviation industry projected a doubling in demand by 2037. Sounds great if you run an airline—but it's a disaster for the climate. Aviation is extremely carbon intensive and very hard to decarbonize. Keeping planes aloft takes a lot of dense energy, and right now that means fossil fuels. There are no viable technical solutions to decarbonize long-distance aviation at scale in the near term. If business-as-usual aviation continues, international air travel alone will account for 22 percent of CO_2 emissions by 2050, which would blow our carbon budget. The math is clear: To stabilize the climate, aviation will have to be decreased, especially by frequent flyers. Reducing unnecessary travel is the first key step. A recent study of frequent flyers found the travelers themselves rated only 58 percent of their trips "important" or "very important."

Reducing frequent flying can be achieved by a combination of carrots (like building clean infrastructure such as high-speed rail networks, which could substitute for many trips by air) and sticks, to make air travel prices reflect their true climate cost. Aviation is currently heavily subsidized, along the whole value chain from suppliers and manufacturers to airports to airlines. Removing these subsidies, for example by ending the loophole that exempts aviation from fuel taxes (which a European Citizens' Initiative aims to do), is a step in the right direction. Ticket prices also need to reflect the climate cost. One option is through increased flight taxes. Another option is a frequent-flyer levy, with sharply increasing prices after the first flight per year, which is perceived as fairer and receives more public support in polling than tax options.

For universities, and for many businesses with a big travel culture, flying is one of the largest parts of an organization's carbon footprint. A recent study found that flying at the University of British Columbia was equivalent to more than two-thirds of emissions from operating the campus. Inspired by the leadership of the Tyndall Centre, after hard work by two of my Lund colleagues, my department has adopted a new travel policy where we aim to substantially reduce our flying for work. After sharing our policy online, we've been contacted by academics from Harvard, Leeds, and elsewhere around the world who are looking to follow suit. We've become connected with networks of people working to get academic societies, conferences, and whole universities to reduce their emissions, declare a climate emergency, and/or adopt carbon budgets in line with Paris; there are far more than I ever realized. And in the wake of the coronavirus, as travel virtually ceased overnight and businesses were forced to come up with new ways to stay connected, major conferences (the largest source of flight emissions for universities, and common in sectors such as medicine) have successfully gone virtual.

Car-Free Cities

Where we live in relation to where we need to be—for work, school, family, and all the other puzzle pieces that constitute our lives—determines so many of the daily patterns, habits, and choices that follow. Living close to work and school is a huge time, stress, health, and carbon saver. In grad school in California, I lived in the country house my ex had always wanted. It was beautiful, but I was entirely dependent on a car to get groceries or see friends. Getting to work by public transportation would have involved getting to a bus stop five miles away, then taking two buses, a ferry, another bus, a train, and a shuttle. The trip was five hours each way; not a very practical daily option. With San Francisco traffic between my home and the university, my typical drive to work was a still-not-great two hours each way. It was not unusual for me to eat both breakfast and dinner in my car.

When I moved to Sweden and got divorced, living alone for the first time made me realize how much *who* I shared (or, by this point, who I *didn't* share) a household with determined so much of how I lived—both in practical ways and with regard to the values that guided my household choices. I sold both (!) of the cars I'd owned in California. In Lund, I chose an apartment in the center of town, within walking distance to everything. In charge of all my own meals, I stopped buying meat.

These more climate-friendly options were available to me through a combination of privilege and societal infrastructure. I had the privilege to choose where I would move because I had a job that could pay the rent, and it came at a reflective time in my life when I was rethinking my values across the board. But even though I was in a position to make these decisions, I could still only choose from the options that were available. When I moved to Sweden, my default options were a largely decarbonized energy grid; plentiful buses and trains, and a society where I felt safe using them alone at night; and dense, well-built, more compact housing. Obviously, personal choice exists in a social context (although privilege often expands available choices). Moving to Sweden made more sustainable choices much more within my reach, and indeed more likely.

Transportation is the largest source of greenhouse gas emissions in the United States. For the average American household, gasoline for private cars is the largest climate polluter. So far, decarbonization efforts have focused on electricity generation, while global transport still runs nearly 100 percent on fossil fuels. As the transport researcher Giulio Mattioli tells me, "Car use is responsible for a lot of emissions, and you should cut it as much as possible." Long car trips, especially for leisure, cause the most emissions, so ways to replace those long drives need special focus. The most effective way to reduce transport emissions is "modal shift"—to reduce car use (whether it's you or an Uber driver at the wheel) wherever possible, and travel instead by transit, walking, or biking. Walking or biking is also a big win for physical and mental health, and most trips by car in the United States are feasible biking distance: Nearly half are three miles or less, and more than 20 percent are a mile or less. (Be sure, though, not to "reward" your short daily cycling commute with long weekend or holiday drives, which can wipe out all the emissions savings.)

To make transport climate-safe, we simply have to get to the root of the issue and work toward reducing the need for cars as much as possible by putting people, not cars, at the center of towns, cities, and streets where it's feasible and desirable to be car-free. For example, by relocalizing and having dense, walkable neighborhoods with local services available; making it affordable to live near work and school; and redirecting the massive subsidies away from private car use toward effective public transport infrastructure and streets that are safe and pleasant for walking and cycling.

Reducing car numbers will address their inherent inefficiencies. Up to 95 percent of the weight of a car is the vehicle itself, not the passengers and goods its purpose is to move around; that's fuel wasted. The rise of gas-guzzling SUVs is especially worrisome. An International Energy Agency analysis showed their rising emissions have more than canceled out savings from electric cars so far. If SUV drivers were a nation, they would be the seventh-largest source of carbon pollution.

The car-based transport model also supports low-density rural housing. These larger homes with their greater consumption and energy use emit about twice as much greenhouse gases as high-density housing. Reducing cars, which spend 96 percent of their time parked, will also free up a huge amount of land for more healthy and lovely uses than parking and roads. A 2020 study led by Felix Creutzig found that car users took three and a half times more space in Berlin than non–car users. All the ethical principles they examined "suggest that on-street parking for cars is difficult to justify, and that cycling deserves more space."

After reducing car use as much as possible, all cars that remain need to be fossil-free. The "death of the internal combustion engine" predicted by *The Economist* in its August 12, 2017, cover story is still a ways off, but many countries have enacted stop dates for selling fossil fuel cars, and sales of new electric cars are increasing. While electric vehicles today do reduce life-cycle emissions compared with gasoline ones, they're not a panacea. Our 2017 research showed that, taking into account the full life-cycle impact of vehicle production and electricity use, electric cars still caused about half as much climate pollution as typical gasoline cars. Even with longer-term improvements expected in car technology and decarbonized electricity, transport cannot achieve its necessary share of emissions reductions without reducing car use.

Car dependence illustrates the problem with the Exploitation Mindset tendency to focus on purely technical shifts (substitute electric cars for fossil cars, done!) without addressing underlying problems. Past technical innovations have tended to greatly increase use of a technology, and therefore environmental impact, even if they have lower impact per use. Further, there is sometimes a mistaken tendency to think that more tech must mean greater efficiency, but the data often tell a different story. For example, a UC Davis report showed that ride-hailing services like Uber and Lyft have led to increased car use and traffic, more trips being taken overall, and fewer trips taken by walking, biking, or public transport. Similarly, the longer-range dream of self-driving cars might free up the one human they transport to do something else while commuting, but cars enable more sprawl, traffic, local air pollution, and greenhouse gases, whether a human is at the wheel or not. Tech is not always the answer; we have to look at behavior change too, by rethinking how to meet our needs in sustainable ways.

Employers need to think about reducing cars and promoting alternatives too, because employees commuting by car is a big part of many company footprints. Our 2018 study found that offering telecommuting and giving financial incentives for employees not to drive to work (like paying them to give up parking spaces, or paying for transit passes) saved the most emissions of all the interventions we studied.

There are some amazing success stories from cities and neighborhoods that have experimented with going car-free. One visitor to Ghent, Belgium, reported "the air tastes better . . . what stands out most is the silence . . . people turn their streets into sitting rooms and extra gardens," adding that their relatives living in the city "got rid of their car because it became obsolete." Another resident noted that "most shopkeepers are doing great business." Many cities rapidly opened streets for bikes and walking under the coronavirus pandemic, showing how fast these changes can take place. Reduced road vehicles contributed the biggest drop in climate pollution from the first few months of the pandemic.

It's encouraging to know that people-friendly cities can be regenerated from car-centered ones. Copenhagen was overrun by cars in the 1960s. It was grassroots campaigns that changed policies and streets to promote bikes over cars, including taking back parking lots and streets from cars for safe, separated bike lanes. People's behavior followed, with the biking share increasing from 10 percent in 1970 to more than 35 percent of all trips. (That's six times more than America's most bike-friendly city, Portland, which has 6 percent trips by bike.) Copenhagen is now touted as one of the world's best biking cities; in 2016, for the first time, more bikes than cars crossed the city. The city calculates that each kilometer cycled earns society $0.74, while each kilometer driven causes a loss of $0.81.

Electrify Everything

Ultimately, to have transportation that doesn't wreck the climate, it will need to run on carbon-free energy, most of which will be electric. Decarbonizing electricity and heat production is the first necessary step of decarbonizing the whole economy; currently the sector is heavily fossil-based, making it responsible for 25 percent of total greenhouse gases globally. Once we have clean electricity, the way to stop climate pollution overall is to electrify everything to run on it, while shutting down fossil fuels.

We already have the tech to make carbon-free electricity, but we need to scale it up massively. As Leah Stokes from the University of California, Santa Barbara has pointed out, we actually need not 100 percent clean electricity, but 200 percent clean electricity by 2035, because we'll need enough clean power to run everything we now run on fossil-based electricity, as well as directly on fossil fuels like diesel, oil, and jet fuel.

Electricity, though, is one area where Seth and I found individual actions were not always the most effective way to go. In the best case, switching to buy green energy (changing your home electricity provider to one running on 100 percent clean energy) saved on average as much as skipping more than seven months of driving. We didn't recommend it as a universal high-impact action, though, because some studies showed problems with double-counting in some European green energy markets (meaning it did not lead to real benefits where newly built clean energy replaced dirty). Another study found installing rooftop solar panels to provide your own electricity at home had high but widely varying potential to reduce emissions. Research by Charlie Wilson and colleagues has shown these smaller-scale solutions can drive faster decarbonization than cleaning up "lumpy" centralized systems.

What has proven effective at decarbonizing electricity is regulation and standards to provide incentives and targets for clean energy supply and demand-side efficiency. The most effective policies to date for reducing greenhouse gas emissions in Europe have been regulations establishing targets and policies for renewable energy (setting dates for an increasing percentage of energy coming from clean sources) and establishing efficiency standards for energy generation, transmission, distribution, and consumption in appliances and buildings.

Personally, knowing that home energy use is the second-biggest source of emissions for the average American, I've decided to reduce my home energy use as much as possible anyway (which also saves money). If I owned my own house, I would look at getting solar panels on my roof to provide my electricity. As it is, I opted for a 100 percent renewable provider, which is increasingly price-competitive. (I do have confidence in the one I chose, but this is a low-risk strategy anyway.) The biggest home energy use is to heat or cool things (including the air temperature in your house, water for washing yourself and your clothes and dishes, and your washer and dryer). This makes big appliances like water heaters the exception to the usual low-carbon rule of thumb to use what you already have, because newer, highly efficient models save a lot of emissions and are worth the upgrade if you can afford it.

After switching to clean energy, the two actions that save the most home energy are "refurbishment and renovation" (like improving insulation and installing efficient windows) and electric heat pumps (basically, reversible air conditioners that can both heat and cool your home by moving air around), according to a 2020 study led by Diana Ivanova.

For my home energy use, I'm limited in making big appliance changes since I live in a rental apartment. I can't replace our inefficient old water heater with a solar one, or heat our home with an electric heat pump, but the ancient heating barely works anyway, so Simon and I are saving money and carbon when we bundle up in the winter. Some of our biggest energy savings at home come from the appliances we live *without*: We don't have an air conditioner, which is the biggest home energy user for homes that have one. (Also, if not properly maintained and disposed of, air conditioners can leak a greenhouse gas up to nine thousand times more potent than CO_2.) We don't use a clothes dryer (hang drying saves a lot of energy and makes clothes last longer); we also wash our clothes in cold water (works just as well with detergents nowadays, and saves heating). For a long time, I found I wasn't using the food in my freezer effectively, so I unplugged it and was using it as an insulated wine cellar.

Most people have the wrong idea about saving energy at home. In a study led by Shahzeen Attari, the top answer given for the most effective way to conserve energy was turning off lights. In fact, replacing light bulbs with efficient ones like LEDs and turning them off when not in use actually saves much less energy than the actions above. I've done it anyway, but it's not where your carbon zealotry should go!

The Future Is Closer than You Think

It can be hard to imagine transforming our transport and energy systems. Indeed, most research focuses on the unsustainable systems we currently have, instead of the sustainable ones we desperately need. In a 2018 study led by Seth Wynes, we screened 2,157 studies of behavioral interventions in high-impact climate domains. We found exactly zero that examined reducing flying. It seems it simply had not occurred to anyone to question or try to change business as usual, or at least to study it when it happened. But when we look around, we can see seeds of change already growing around the world.

For example, in the last few years in Sweden, avoiding and reducing flying are rising up the social, political, cultural, and economic agenda. In 2016, the popular opera singer Malena Ernman publicly declared she quit flying for the climate. The term *flygskam* (flight shame) and a Swedish hashtag "I stay on the ground" started circulating in Swedish media in 2017.

In January 2018, the journalist Jens Liljestrand wrote a heartbreaking personal account in a Swedish newspaper titled I'M TIRED OF SHOWING MY CHILD A DYING WORLD. He described taking his family diving in Kenya and finding only a reef full of slime. In the car ride back to the hotel, his eleven-year-old daughter broke the silence: "There were no colors. In all the movies it looks so pretty with all the colors. But now I know that's just in the movies. It's never like that in real life." Liljestrand declared he was done flying to "take selfies with a dying world in the background." His story accelerated the debate, and more publicly joined him.

The Swedish Green Party proposed a flight tax, though it was set at a lower price than research showed consumers were willing to pay. Still, flying policy became an agenda issue for the 2018 election. Ensuing debates raised the issue that jet fuel (kerosene) is not taxed due to a 1940s loophole, generating broad support across political parties in Sweden for the idea that "flight should pay its climate cost," though each had different proposals for how to design the policy. The Swedish municipality of Helsingborg added a 50 percent fee on business flights for city employees, which was used to pay for bike lanes and public transport.

Meanwhile, research was making flying's carbon footprint harder to ignore: A 2019 analysis showed international flights taken by Swedes (which are not counted under the country's emission reduction plans) produced as much climate pollution as the largest national source of pollution, private cars. In Europe, the low-cost airline Ryanair joined the top ten climate polluters in the EU, alongside nine coal-fired power plants.

While many airlines focused on greenwashing, including promoting carbon-offset schemes shown not to reduce emissions, at least some in the airline industry were more forthcoming about the climate harm of flying. The Dutch flight company KLM produced an ad encouraging customers to "fly responsibly" and asking "Do you always have to meet face-to-face? Could you take the train instead?"

Climate change was also making warming harder to ignore: A brutal drought in Sweden in summer 2018, including months without rain, devastated crop yields. Farmers with no hay to give their animals were forced to slaughter them early. Sweden is well prepared for snow and ice, but not for heat extremes. Forest fires raged north of the Arctic Circle. Countries as distant as Italy answered Sweden's call for help with firefighting resources.

While on parental leave, Maja Rosén launched the "Flight Free 2019" campaign in November 2018. Her goal was to attract one hundred thousand people (one out of every hundred Swedes) to pledge to stay on the ground for 2019, to take both individual and collective action. She has been an effective, powerful, and empathetic voice in drawing people into conversation with her message that "people care and that most people would be prepared to fight for the climate if they realised the severity of the situation and the importance of their own actions."

Social media also got involved. Membership exploded in Facebook groups like Tågsemester (Train Vacation) to discuss and share tips on train travel, and a network called Sustainable Influencers used their social media platforms to make low-carbon lifestyle choices—including local tourism, plant-based diets, secondhand and vintage clothing, and ways to go plastic-free—look sexy, fun, and desirable. A Swedish newspaper chartered a train from Stockholm to Venice in summer 2019, which was filled with journalists and others blogging about the experience.

Of course, there was also the anonymous Instagram account @aningslosainfluencers (@cluelessinfluencers), which problematized the usually glorified high-carbon lifestyles of celebrities, superimposing labels like "0.9 tons CO_2 per person" on a picture of a weekend getaway to Barcelona, and "57 years of sustainable flying in 15 months!" calculated from the flight-bragging posts of a woman shown posing in front of palm trees. The account helped take the *flygskam* debate to international coverage in summer 2019 in outlets including *The Guardian,* Bloomberg, and *The Wall Street Journal.* This phenomenon has sometimes been misinterpreted as "flight sham*ing,*" finger-pointing at others. In fact, research showed that those who changed their behavior did so more from their own desire to live in accordance with their values than because of being shamed by others.

The cultural change, from conversations in the media and over lunch tables, started to take effect. Books, articles, and hashtags celebrated Swemester—local holidays. In 2019, domestic train travel was up 11 percent. Meanwhile, domestic flights were down 9 percent, and international flights also decreased slightly. A survey found that 14 percent of Swedes avoided flying in 2019 because of the climate.

What most astonishes me about this transformation underway is how few people it took to get it started. Recent research shows that contrary to the expectation that a majority must support change, only 25 percent of a population is actually needed to support a norm change, like not smoking or not flying, before it reaches a tipping point and becomes a social norm, and this can happen fast. And a few individuals can play a huge role. Maja Rosén has had an incredible impact in shaping national and international conversations through the media and inspiring other groups to start in countries including the United States, the United Kingdom, Canada, Australia, and France—all in the first year of her campaign, despite signing up less than 15 percent of her target audience.

Greta Thunberg has had astonishing international impact,

drawing global coverage for her zero-emission sail across the Atlantic in summer 2019; but before the world had heard of her, she
was the inspiration for her mother, Malena Ernman, to stop flying and helped set off the cultural change underway in Sweden.
These seemingly small changes, starting with one person taking
a stand, sharing their story, and rippling out to others, contribute
to the cultural narrative, shift how people see themselves and
what they expect of others, and ultimately shape what is possible.

Low-Carbon Love

My frequent-flyer gold card, hard-won but short-lived status symbol, expired in 2012. It's now on display in a museum exhibit called
Carbon Ruins, which looks back from 2053 at bizarre artifacts
from the recently concluded fossil age, alongside fast-food burgers
and petrochemical plastics.

When I started dating after my divorce, the difficulties of rec-
onciling romance and my climate ethics were quickly apparent.
One of my first dates was with a dude whose online profile tagline
was "Elegance and charm" (barf). He mentioned that he owned a
home in a rich coastal town. This was meant to be a marker of his
sophistication and class. All I could think about was how his property was slated to succumb to the rising seas, a fact this suave gent
was not mentioning. We didn't have a second date.

After six years of intermittent online dating, I decided to try a
new tack: honesty. I updated my Tinder profile to begin with a list
of the qualities that previous partners found unattractive, to filter
out the nonstarters earlier: "Energetic, impatient, quixotic, and
ambitious brainiac, dilettante, and closet romantic seeks brutally
honest, wickedly funny, kind-hearted nerd and/or creative type."

A month later, I met Simon. On our first date, we talked about whether math was discovered or invented, and our respective pending astronaut applications (sadly NASA didn't want me, and the Canadian Space Agency didn't want him). I thought the too-good-to-be-true veneer of this guy would wear off on the fifteen-hour train trip we took to Paris for our fourth date, but we liked each other better by the time we pulled into Gare du Nord. (I highly recommend a long train trip as a relationship crucible!)

Two years later, as we planned our wedding, it felt natural to combine our love for each other, our families and friends, and adventurous slow travel. After a ceremony with our parents and siblings, we took a train trip across North America to celebrate with friends and family in regional gatherings hosted in friends' homes, in a sunny park, by a campfire, and on the shore of Fallen Leaf Lake. Instead of having two hundred people fly to us to re-peat the same conversation in five minutes of face time with each of them, we had quality time with intimate groups of people spread out over three weeks. Deliberately slowing down gave us more time for what matters.

Now we have another low-carbon adventure in mind: a goal to sail from Sweden to North America. We took our first sailing course together in the summer of 2019 under the midnight sun in Norway.

Chapter 10

Food Shouldn't Come from a Factory

Putting Grandpa George's Turkey out to Pasture

Confession time again: Reader, I am a turkey heiress. A turkey heiress with a problematic family legacy. Here I want to use another ancestor's story—that of my paternal grandfather—to illustrate that the idea that every human life literally outlasts itself for centuries is true for the traces we leave on the land as well as in the atmosphere.

My grandfather George Nicholas helped invent the system of treating animals like an assembly-line product—specifically, he figured out how to breed turkeys and turkey parts for optimized, scalable production and consumption. He was a product of his time and of the many forces that pushed toward producing ever more as the ultimate aim; he was hailed as a pioneer.

But much of my life's work has been studying the overwhelming harm today's system of industrial agriculture causes life on Earth. The food we produce and consume—what, how, and where—needs to change fundamentally. Those of us alive today know that deep, regenerative growth does not happen on an assembly line, and it's on us to build systems that cause regeneration, not damage, in the

173

long term. It's time to recognize and step back from the assembly-line values and expectations ingrained in our food system.

The Problem with Bigger Breasts

In 1939, my father's parents received a loan from a program earmarked for young couples in California to take over farms and ranches abandoned or foreclosed upon in the Depression and make them productive again. Grandpa George was famously cheap. At the poultry-processing plant in Fulton, he hated paying workers a quarter an hour to pluck the tiny but unsightly dark pinfeathers from the dead birds before they were sold on supermarket shelves. On the advice of a professor at the University of California, Davis, in the 1950s my grandfather experimented with turkey breeding to lighten those pesky dark feathers.

The resulting turkey had a Thanksgiving-ready breast so large that the birds could not reproduce naturally. Therefore, much of the business of Nicholas Turkey was masturbating male turkeys to obtain turkey semen with which to artificially inseminate the females and produce fertile turkey eggs. These eggs were shipped to their clients around the world, who would raise them, usually packed indoors. With optimized feed (a mix of grains like corn, soy, wheat, and barley, plus canola oil), you could go from hatchling to a market-ready assembly of legs, thighs, and breast in as little as fourteen weeks. The Nicholas Broad Breasted White would go on to become the world's main turkey.

If you eat a supermarket turkey for Thanksgiving, my butterballed brethren, you're eating Grandpa George's turkey.

This piece of family folklore is a microcosm of the industrial, homogenous, consolidated food system that emerged through the twentieth century. The Exploitation Mindset focus was on maximizing food production, which was spectacularly successful. But this narrow focus came at a very high price to human health and now threatens the health of the whole planet.

Together with colleagues from three Swedish universities, I analyzed more than twelve thousand scientific papers from the last fifty years on how to feed the world. We found that increasing production through technology became the main focus (about 75 percent of studies by 2018), while attention to what we eat (dietary choices) remained consistently low, hovering near 25 percent. Indeed, while low yields remain a problem for the poorest farmers, in many rich countries the problem is overproduction of unhealthy food using unsustainable methods, where cheap prices don't include the cost to health and the environment. It's time to apply a little Regeneration thinking to get to the root of the issue of how we use land and find ways to do so that meet human needs for food while protecting ecosystems and increasing resilience.

Fragile and Inefficient

Today's food system is precarious because it is homogenized and overspecialized instead of diverse and resilient. The Green Revolution tripled crop yields by reengineering the planet's flow of nutrients to maximize the size and quantity of three species of grain: rice, wheat, and maize (corn). Today, two-thirds of calories that humans consume come from just these three plants. This means people are increasingly pinning our survival to a narrowly optimized, and therefore highly risky, food production strategy. Relying on just a few species and breeds to feed so much of humanity means we're increasingly vulnerable to a single disease, or a change in weather conditions (ahem), wiping out a food source. The near-total dominance of Grandpa George's turkey is a prime example of this problem of overspecialization.

Trade can help bolster food security for everyone under extreme weather, and especially for import-dependent regions in North Africa, the Middle East, and the Andes. But today's food system is made up of highly specialized, complex, and distant supply chains, which are vulnerable to shocks and disruptions. A 2013 study led by Marianela Fader found that sixty-six countries currently lack the ability to be self-sufficient for the diet they consume today, due to land and water constraints; 16 percent of the world population relies on international trade to meet their food demand.

The majority of harm from our food system today is caused by livestock production. A comprehensive 2018 study in *Science* found that animal agriculture occupies 83 percent of farmland, produces 58 percent of agriculture's greenhouse gas emissions, 57 percent of its water pollution, and 56 percent of its air pollution, and uses more than a third of the fresh water withdrawn for agriculture. Even though animal agriculture creates most of the damage, it provides comparatively little nutrition: just 37 percent of protein and 18 percent of calories.

In sustainable food systems, animals turn waste into resources, but in today's industrial food systems, animals turn resources into waste. Under regenerative agriculture, there could be enough animals around to eat waste that people can't (like grass or inedible leftovers) and turn it into valuable resources, like fertilizer (animal poop), food (meat, milk, eggs), and materials (leather, wool). Industrial farms do just the opposite: They feed valuable resources like grains and clean air and water to huge numbers of animals concentrated in a tiny area, then release greenhouse gases and polluted air and water.

Industrial farms are designed for producing livestock efficiently, but there's one inherent inefficiency they can never overcome: the energy lost as you go up the ecological pyramid. The job description for being a plant is converting sunlight into biomass, while the animals that people eat need to eat plants to grow. Instead of feeding leftovers to animals, we now fatten them quickly on food that people could eat, like soy and maize. An analysis led by Emily Cassidy found that more than a third of food grown went to feed animals. This is an inefficient detour, especially for beef cows, which convert only 3 percent of the calories and 5 percent of the protein in the plants they eat into food for people. The authors concluded we could feed an additional 4 *billion* people (about half as many as are alive today) if we fed that food directly to people. Another study found that if the whole world ate the average Indian diet, we could feed the world on less than half of today's cropland. Imagine if we stopped expanding cropland, and in fact shrank the footprint of cropland, while feeding people healthier diets.

Industrial Agriculture Is Killing the Planet

The way we produce food today is killing life on Earth and the conditions necessary to support it. We are rapidly depleting and disrupting the healthy soils, intact food webs, and natural cycles of water and waste recycling that underlie and support humans' ability to grow food. The planet is already breaking down under the level of industrial farming we have today.

Humans now dominate planet Earth; our use of most of the ice-free land on the planet leaves only 28 percent for the rest of nature. (Of these areas with "minimal human use," just over half are forests and other ecosystems, while the remainder is classified as "barren, rock"—you can just hear nature saying, "Gee, thanks.") Currently, humans use about half of all ice-free land to feed ourselves: an area the size of South America (12 percent of ice-free land) to grow crops, an area the size of Africa (21 percent) to permanently graze livestock, and an area the size of North America (16 percent) for occasional grazing as well as hunting and gathering wild products. We dominate most forests on Earth; 22 percent of ice-free land is covered by forests managed as plantations or for timber or other uses, more than twice as much as remains in intact or primary forests. The direct footprint of cities, roads, and other human infrastructure is just 1 percent of land, tiny compared with the vast swaths of countryside we draw from to feed our cities.

Agriculture as practiced today is the main driver of the biodiversity crisis. Humans are causing extinctions *one thousand times* faster than their natural background rate; up to 1 million species are now threatened with extinction. We are hacking down the tree of life by replacing forests with farmland (the cause of 80 percent of global deforestation, including in biodiversity-rich tropical areas). Most Amazon deforestation is to feed animals, either directly through pasture or indirectly by growing soy as animal feed. Richard Waite, a sustainable food expert with the World Resources Institute, advises omnivores concerned about soy and deforestation to eat *more* soy (like tofu) and less meat from animals that are fed soy, because plant-based diets reduce this pressure for land. When forests and grasslands are cleared and plowed, their inhabitants have nowhere to live.

This habitat destruction and degradation are driving a catastrophic collapse of life. The 2020 *Living Planet Report* finds that population sizes for wild vertebrates have declined an average of 68 percent since 1970. The decline in freshwater populations, including frogs, turtles, large fishes, and critters like river dolphins and beavers, is 84 percent. Eighty-four percent. I teared up writing that sentence.

It keeps going: globally, 87 percent of the water that people consume goes to irrigated agriculture, which uses twenty-two times more water than the households of 7.8 billion people. (This means switching to a more water-efficient diet, like from beef to bean burgers, saves *way* more water than taking shorter showers.) Fresh water is an incredibly limited resource. The US Geological Survey visualizes what it would look like if you collected all the Earth's surface fresh water from lakes and rivers: a bubble only thirty-five miles in diameter. In an afternoon bike ride, you could traverse that bubble of water that people and every other living creature need every day to survive.

Water overuse from agriculture literally erased what was previously the fourth-largest lake off the map. In the 1960s, the Soviet Union diverted rivers that formerly fed the Aral Sea, which spanned the borders of Kazakhstan and Uzbekistan, to grow cotton in the nearby desert. With more water evaporating than was replaced by inflow, the lake began to shrink dramatically, splitting into two lobes. By the year 2000, it resembled a pair of lungs thinly connected at the top. That connection shriveled, along with nearly all of the eastern lobe, which disappeared entirely in 2014 (though it has returned in heavier rainfall years since). Today the Aral Sea is a scattering of briny pools covering about 10 percent of the original lake bed. Rusted boats from former fishing communities are stranded in a desert, miles from the nearest water.

Taken together, how humans use land causes almost a quarter of all climate pollution, making agriculture and forestry second only to fossil fuels as a driver of the climate crisis. Some of this climate impact is from carbon released from plowing and deforestation, but most of agriculture's contribution to climate warming comes from two shorter-lived but more powerful greenhouse gases, methane and nitrous oxide, where agriculture is the dominant global source. Cows and sheep emit methane (mostly through burps, not farts!) when microbes break down grass in their guts. Nitrous oxide is created from applying nitrogen fertilizers, often in excess.

The chemical dependency of industrial agriculture contributes to its high carbon footprint and poisons wildlife. Industrial agriculture sprays fossil-derived chemical poisons on our food to kill the bugs who'd like to eat it first. Some of them die, along with many of the natural predators who would help keep them in check; these innocent bystander insects are followed by the birds who eat them. Germany's alarming 78 percent decrease in insect abundance over just one decade and the 53 percent decline in US grassland songbird populations since 1970 have been linked to industrial agriculture, including more toxic pesticides. This is especially scary considering one-third of food production globally depends on animal pollinators, including a multitude of key crops that depend on insect pollinators, like almonds, apples, cherries, cocoa, and coffee. Meanwhile, pests and weeds rapidly evolve resistance to outsmart our chemical cocktails, creating a treadmill of dependence on ever more toxic sprays.

Nutrients like nitrogen and phosphorus are essential for plants to grow, and thus for people to have food. Artificial nitrogen fertilizer is created using a lot of energy (hello, fossil fuels!) to extract nitrogen from air, where it is abundant. But phosphorus is limited. After a brief glory day for the guano (bird and bat shit) industry as a major global source, today most phosphorus is dug from the ground, mostly in Morocco and its disputed territory of Western Sahara. These geological phosphate reserves are replenished at the speed new rocks are made. Humans are mining them way faster than that, depleting reserves. Digging up what remains is becoming ever more expensive and increasingly out of reach for poor farmers, threatening world food security.

While some areas, particularly in Africa, remain underfertilized, globally we have too much of a good thing: Both nitrogen and phosphorus are dangerously out of balance. Too much phosphorus and nitrogen is applied to fertilize cropland; the sewage we flush down the toilets and manure from our livestock also add excessive nitrogen to the system. These nutrients run off in rainwater, creating belly-up fish in lakes and dead zones of algae blooms in places like the Baltic Sea and the Gulf of Mexico. An international team of scientists led by Will Steffen estimated that global phosphorus runoff to the oceans is twice the sustainable level, and nitrogen is nearly three times the sustainable level. Agriculture is the largest cause of algal blooms.

At least we know where to focus on more sustainable farming interventions to help bring the global biogeochemical cycles in balance: Most of the excessive fertilizer application and runoff is in just four intensive farming regions: the US Midwest, Western Europe, China, and northern India.

Industrial Agriculture Is Killing Us

The planet is suffering from our current food system, but so is our health. Globally, nearly three times as many people today suffer from the malnutrition of excessive, unhealthy food than from the malnutrition of going hungry. Medicine and public health have been so successful at treating infectious disease and improving sanitation that unhealthy diet was the largest mortality risk factor globally in 2017, linked with one-fifth of adult deaths (11 million deaths). (For comparison, tobacco use was second, linked with 8 million deaths; the terrible coronavirus pandemic had caused 1.8 million confirmed deaths as of December 2020.) A shift to healthy and sustainable diets would be a major win both for people and planet, with plant-based diets shown to reduce type 2 diabetes more than 40 percent and death from heart disease by 20 percent.

Not only is today's food system cruel and unhealthy; it endangers our ability to fight disease. This issue hits close to home for me: My father made his lifelong career in industrial agriculture, but the system of factory farming is now putting his health at risk. He has an inoperable spinal infection that requires lifelong antibiotics to manage, just as antibiotic resistance is on the rise. The largest source of antibiotic-resistant bacteria may be farm animals, which are overfed antibiotics to keep them alive and growing rapidly in unhealthy conditions. We need more holistic approaches to simultaneously address human, animal, and environmental health, such as the rising field of One Health, which convenes experts from all three disciplines.

It was a fundamental Exploitation Mindset mistake to treat animals as an assembly-line product to be optimized by producing as many, as fast, and as cheaply as possible. Thinking about nature in this way, particularly thinking about other sentient creatures in terms of the "lowest cost of liveweight production" and "least cost deboned breast meat," as one turkey industry guideline calls it, illustrates the thinking that has gotten us into the ecological crisis we now face. It's wrong to raise animals in conditions where they suffer. For the Nicholas Broad Breasted White turkey, not only do the poor birds never get to have sex, but their legs often struggle to support the weight of their massive breasts. Treating life as just a process to be broken down into component parts at the cheapest possible price (semen/egg/turkey; breast/thigh/leg), we end up with the unacceptable system of factory farming we have today, which endangers both human and planetary health.

Regenerating Nature

Industrial agriculture illustrates how our current ways of exploiting land break and disrupt natural cycles, fraying the web of life. We need nothing less than the Regeneration Mindset for people to find ways to work with rather than against nature, including in our interactions with the land.

This shift is already starting to happen. After generations of seeing nature as something to be controlled and managed toward some "optimal" productive end, some land stewards are developing a new aim: rewilding. Ecologist Tobias Kuemmerle describes rewilding as promoting complex, self-regulating ecosystems with their integrity restored, where nature takes its course. People are increasingly recognizing that land use dominated by industrial agriculture makes us vulnerable, as does replacing natural disturbance regimes (like periodic forest fires or flooding of meandering river valleys) with overmanaged, sped-up, highly disturbed and simplified systems more prone to disaster. It's a big step toward regeneration when people practice adaptive management (learning by doing with continual assessment), and shift the focus from "who's there" (the identity of species present) toward supporting system function, including replacing missing or degraded components. A popular example is the reintroduction of wolves in Yellowstone National Park, which has helped bring back beavers, songbirds, and forests.

Protecting and restoring wildlife are important, because animals going about their business of being animals as part of intact ecosystems does a great deal to keep the planet healthy and habitable. Chris Doughty and colleagues found that today's biologically impoverished ecosystems have lost 94 percent of their capacity to move nutrients through animals across land and sea. That means that today, valuable nutrients that big mammals used to eat, recycle, and disperse across land and the surface oceans now wind up, unusable, on the bottom of the seafloor. For example, whales dive deep in the ocean to feed, bringing the nutrients contained in the small, deep-dwelling fish up to the surface with them and releasing them through the majestically named "fecal plumes." (I can highly recommend two minutes of entertainment listening to the diver Keri Wilk describing being caught in a pungent whale "poonado.") Saltwater fish that swim up freshwater rivers to breed, like salmon, bring ocean phosphorus with them that predators like bears, eagles, and otters distribute on land. But today, salmon populations in the Pacific Northwest have declined more than 90 percent, and whale densities have decreased 66–99 percent. Doughty suggests that restoring animal populations could effectively recycle nutrients now being lost to sea; plus, how cool would it be to live in "a world of giants"?!

Personally, we can help regenerate nature in our backyards in many ways, from choosing native plants to providing water and habitat for wildlife. (The best thing cat owners can do is keep those deadly hunters inside. If they must go out, put a goofy-looking colorful clown collar on them to warn birds, and make sure they're spayed or neutered; those adorable kitties and their feral cousins are likely the largest human-caused threat to bird and mammal populations in the United States.)

The biogeochemist Geneviève Metson has lots of ideas for replacing industrial animal agriculture with regenerative natural cycles, including fertilizing crops with the nutrients we now throw and flush away in food, human, and animal waste (yes, pee and poop!). Whereas the Exploitation approach is to increase nutrient supply through more extraction using increasingly fanciful tech solutions, she takes a holistic focus to look at how sustainable diets could decrease nutrient demand and how recycled nutrients could meet the demand that remains. With colleagues, she has calculated that there is just about enough phosphorus in human and animal waste in Sweden to meet crop demands; more than half of crop fertilizer needs could be met by waste generated within a few kilometers, which is important because waste is heavy and expensive to transport. In Sydney, Australia, she finds there is fifteen times more phosphorus currently going to waste than is demanded by local crops. Efficient systems already exist to recycle waste safely, like the ecological sanitation approach used since 2006 by the organization SOIL in Haiti. The key will be minimizing transport distances, and also developing social acceptance to overcome the yuck factor and embrace a more diverse, healthy, and flexible food system.

Eating on a Living Planet

How can we work toward regeneration in agriculture and on our dinner plates? At the system level, we need to rethink the purpose: move away from today's Exploitation Mindset goal to produce ever more of a few food items, blind to their true costs, and toward a food system that produces healthy, diverse, nutritious food, using methods that protect human health and biodiversity. Project Drawdown finds that a range of improved land and agriculture techniques has enormous potential to reduce emissions, store carbon, and improve livelihoods and health, including regenerative agricultural practices like crop diversity to build soil health.

To start growing a sustainable food system, we'll have to redirect the agricultural subsidies supporting today's unsustainable one, which currently total $1 million *every minute* globally. A small, powerful group benefits from the current system; in the European Union, more than 80 percent of farm subsidies go to just 20 percent of farmers. Our 2020 study led by Murray Scown found that nearly half of EU farm subsidies were misspent: increasing income inequality, supporting the most climate-polluting farms with the lowest value for biodiversity, and diverting rural development funds to cities. Incentives need to change so that public money is supporting a healthy, fair, and sustainable food system and so food prices reflect their true health and climate costs.

As with climate stabilization, there is no single remedy to repair humanity's broken relationship with nature, and it won't be a quick fix. A 2020 *Nature* study found that it will take immediate ramp-up of conservation, restoration, planning, and food system efforts "of unprecedented ambition and coordination" to start turning around the screaming plummet of biodiversity over the next several decades—but it's possible. Similarly, a 2018 study showed that it will take a combination of measures to create a food system compatible with a living planet, from dietary changes to technology to eliminating waste.

That said, the single most important choice, both personally and systemically, to start moving our relationship with land and nature toward regeneration instead of exploitation is a shift toward more plant-based diets. Our research showed that plant-based diets were the highest-impact personal climate action when it comes to food. A global analysis by the World Resources Institute found that a menu of twenty-two solutions was needed to achieve a sustainable food system; shifting to plant-based diets was the most effective. The personal dietary change with the single biggest environmental benefit is cutting beef, which uses far more resources and produces far more pollution than other options. Analyses have shown that we simply cannot meet even the 2°C goal without reducing today's meat consumption. Like continuing fossil fuel use, our continued overconsumption of animals is putting off what's necessary if we want a stable climate. Generally, even the best-raised animals have a far higher land, water, and climate footprint than even the worst plant-based options. The majority of climate impact comes from what you eat, rather than where it comes from (transport is usually a small fraction of the total emissions).

If eating animals is part of a regenerative future, the livestock that remain will live on leftovers, not on food that people could eat. They are certain to be a much smaller part of a healthy and sustainable diet than what we eat in wealthier countries today. A recent study calculated that a sustainable food system of "livestock on leftovers" (where animals only eat leftovers humans can't, or graze areas where it's helpful for biodiversity) requires 80 percent less livestock than raised today in the Nordic countries studied. In other words, we currently raise five times too many animals to balance resources and waste.

We need to think about animal welfare too; I don't consider it a win for sustainability if animal production shifts to battery cage chickens, an efficient but inhumane system. Factory farms should be put out to pasture and replaced by humane livestock raising. Michael Pollan writes about one such farm, where each pig is given the "opportunity to express its creaturely character—its essential pigness," to oink and root its way through scraps and generally be just as happy as a pig in shit should be.

The EAT-*Lancet* global analysis identified a healthy and sustainable diet that could feed the 10 billion people expected to live on Earth by 2050 within planetary limits and with greatly improved global health from today. The average American consumes twice as much protein as recommended for health; the EAT-*Lancet* authors found North American red meat consumption was six times higher than is healthy and sustainable, and egg and poultry consumption was about 2.5 times too high. Meanwhile, we should be eating about twice as much fruits and vegetables (to cover half our plate at the average meal), and five times more legumes (like lentils and peas) and nuts. (Don't listen to the Almond Police. Vegan milks illustrate the principle that plants are basically always a better deal for the planet than animals. For example, cow milk uses about twice as much water and emits five times as much climate pollution as almond milk, the climate winner; other vegan milks use even less water.)

The Harvard-recommended "Planetary Health Plate" includes no more than the equivalent of one eight-ounce glass of milk (or one ounce of cheese, about the size of your thumb) per day; two eggs per week; two servings each of fish and chicken per week (each about the size of a deck of playing cards); and two quarter-pounder beef burgers per month. To minimize the climate and biodiversity impacts of fish, *Grist*'s Eve Andrews suggests "mollusks are mighty" (oysters and mussels clean the water they grow in), "feeder fish are fine" (small fish like sardines, herring, mackerel, and anchovies are less overfished), and "big fish are big treats" (at the edge of the plate, wild-caught or certified sustainable, preferably frozen so it doesn't have to be flown).

In terms of food impacts on biodiversity, a 2020 Swedish study found high impacts from meat, especially lamb, but also crops grown in the biodiversity-rich tropics like coffee, cocoa (chocolate), and bananas. In addition to going plant-based, biodiversity-friendly options would be switching from bananas to apples or berries, from coffee to tea, and from chocolate to . . . Well, that nice crisp apple or those juicy berries would be a win for the planet, if not *exactly* replacing chocolate, sorry!

How do I put these dietary choices into practice? In 2014, I accepted a challenge from friends to eat vegan (no animal-based products) for a month. The four of us took turns hosting a weekly dinner and swapped recipes and tips. I tasted my way through the

plant-based milks and butters and found options I liked better than
the cow-based ones. I was surprised how much easier it was than
I'd anticipated to find tasty food (I found Indian and Thai cuisines
especially vegan-friendly). When the month was up, I added back
some occasional cheese, eggs, and fish but didn't miss the rest.
One-month challenges have gotten popular as a way to try out
a sustainable change in a supportive community. My colleague
Karen O'Brien at Oslo University runs cChallenge workshops where
people undertake one month of personal behavior change and
share their experiences of wider change along the way. Everyone
from students to pensioners to the mayor has participated.

Research has shown that restaurants and cafeterias can help
promote plant-based foods to a wider audience by nudges such as
putting them first on the menu and in buffets and emphasizing
their delicious taste rather than harping on their health or environmental benefits (which seems to only preach to the choir).
Given the need to reduce animal consumption up to six times from
today's baseline, it makes sense that many organizations, offices,
and conferences are adopting a default vegetarian menu and supporting caterers who offer tasty plant-based options, as we started
at my department years ago.

One important way to connect with nature is to grow your own
food—even some herbs in a windowsill or a tomato plant on a
balcony. I'm deeply gratified by tending and enjoying peas, toma-
toes, and the fearsome productivity of the squash family in our
allotment garden. Community gardens (and some "guerrilla gardens" in reclaimed parking spaces and median strips) are increasingly popular ways to cultivate both community ties and food.
While people in cities will likely remain reliant on rural regions
to supply some of the food and resources they consume, there is
a lot of untapped potential to produce more food where people
live. For example, a 2014 study found that meeting the vegetable

consumption of city dwellers would take less than 25 percent of the land in most European cities, less than 10 percent in the United States, Germany, and Australia, and was especially feasible in smaller cities, with larger areas per capita. Now, there's an idea for what to do with space freed up from parking! My former student Kyle Clark and I found that planting food trees in cities could provide a great deal of food and a more robust local food system, as well as benefits to wildlife and nature, and potentially social benefits from community engagement and employment in caring for the trees. There are lots of yards as well as unlovely medians, alleys, rooftops, and other corners that could be beautified and brought to life and provide some food in return.

Those of us lucky enough to choose what we eat for three meals a day are voting with our forks for how we treat the natural world. For food security, for the climate, for our health, for the rest of life on Earth: It's time to leave the exploitative food system we have now in the dustbin of history and build a regenerative one.

Chapter 11

Values and Costs

We Can't Afford Not to Fix Climate Change

In October 2017, fire roared toward my beloved hometown of Sonoma, California. My parents left their house in the hills with the clothes they were wearing and an extra pair of underwear they'd grabbed. They first went to my sister's house on the outskirts of town. When the fire crept closer from the wildland edges to suburban cul-de-sacs, my whole family reevacuated to a friend's house twenty miles south in Marin County. I was on the phone with my sister as she used her garden hose, usually reserved for watering her roses, to spray the roof of her house to make it a bit less flammable. Before she hung up, as she packed up her car with her family and their lumbering, ever-cheerful three-legged yellow Lab Ollie, she told me, "We're walking out the door with everything that matters."

My best friend from junior high was caring for her aging parents as their house in downtown Sonoma was enveloped in smoke that hurt to breathe and shortened visibility to less than a city block. Molly's mother had suffered from a degenerative disease for a decade and had recently been put in hospice care. A mandatory evacuation order closed in on their street, but moving her would be medically difficult and emotionally impossible. For a week, Molly and her parents watched as the glow of the fire burned on the surrounding hills and smoke roiled closer.

Even far away from the evacuation zone, instead of "Sunny" or "Cloudy," weather apps reported the status "Smoke." The cloud of ash dispersed, and the charred remains of someone's grandmother's quilt, someone's wedding album, someone's baby shoes, gently settled one particle at a time on SUVs in driveways a hundred miles away.

The American West is wildfire country, but I'd never imagined something the scale of this fire was in the realm of the possible. I could not comprehend the size of the area burned, no matter how many tens of thousands of football fields or what percentage of Rhode Island it was compared to. It felt surreal. There were multiple fires burning simultaneously in places separated by mountain ranges and major highways, at distances that would take hours to drive. As the days went on and the widely scattered fires steadily devoured their fuel, some merged across county lines to become one massive fire.

With a disaster of this scale, the news media was limited in what it could cover, and rumors were flying. Thousands of miles away in Sweden, I was constantly checking in on friends and family and obsessively following social media for news from Sonoma. I saw photos of a high school gym transformed to a refugee camp for the community I grew up in, the floor covered with plastic sheeting, fold-out cots each bearing a neatly shrink-wrapped kit from the Red Cross. I watched videos of people fleeing for their lives along roads engulfed in flames. Evacuees numbly refreshed websites and pored over pixels and wind directions, trying to divine if they would have a home to return to.

On day five, I woke up to the news that the artifacts from the historical landmarks I'd grown up visiting on the Sonoma Plaza were being evacuated. Crews were preparing for the very center, the heart of my town, to burn. I dissolved in tears.

By the time the fire was put out, thanks to hundreds of brave crews who came from all over California, the United States, and the world to work as hard as humanly possible to save all they could, more than five thousand homes had burned to the ground, including those of many friends. The school where my friend Joey taught was gone; his students would spend the rest of the school year in trailers scattered around the city. Whole neighborhoods were now flattened, smoldering ruins.

Dozens of people died in these fires, including an elderly couple too frail to escape their home in time. They had recently celebrated their seventy-fifth wedding anniversary. I realized with a sick dread that they were the grandparents of a girl I knew from high school.

Climate change is killing the California I grew up with and endangering the lives and landscapes I love. The combination of hotter spring and summer temperatures, reduced snowpack and earlier snowmelt, and declines in rainy days during the fire season creates conditions for larger, more destructive fires. The expectation is for nature to absorb carbon, but California's ecosystems since 2001 have acted as carbon sources rather than sinks, releasing more carbon than they absorbed, when fuel buildup from a century of fire-suppression policies collided with increasing wildfires. A 2016 study found that human-caused climate change caused drier fuels and had nearly doubled the area of forest fires in the western United States since 1984. A subsequent study found that hotter, drier fall weather, which bears the fingerprint of human-caused climate change, has more than doubled the days of extreme autumn fire risk in California since my childhood. Exhausted firefighters darkly joke that there is no more fire "season," just a continuous fire year.

The idea that having money makes you safe from climate change is wrong. The devastation from the wine country fires happened in rich neighborhoods, in a rich state, in the richest country on Earth. Climate disaster happened to me. Wherever you live, however much money you have, it's increasingly likely to happen to you, as a climate system fueled by more heat energy creates more chaos, from stronger hurricanes to higher flooding to more intense fires. No one is safe in a destabilized climate.

The California wildfires, and other increasing climate disasters around the world, illustrate some critical points about the costs and benefits of climate action. It's a huge and tragically expensive mistake to think that we can afford *not* to fix the climate. The Exploitation Mindset lies behind many of the flawed economic approaches that are currently delaying climate action and causing increasingly intolerable harm. Our economic task under the Regeneration Mindset is clear: We have to align money, lending, and investment throughout the economy with an intact climate. As always, this involves addressing root causes: divesting our support (financial and otherwise) from systems of climate harm and reinvesting them in sustainable systems, from the global marketplace to your personal finances.

It's Not Just the Economy, Stupid

One clear sign that our economy isn't working as it should to deliver real value is that it's currently cheap to buy dangerous things (fossil fuels, industrial meat) and expensive to buy sustainable things (community renewable energy, sustainably produced food). This means in the last few decades, unsustainable technologies and practices (flights, factory farms) were scaled up because they were artificially cheap, often highly subsidized. High emitters now greatly overconsume these harmful things. It's time we realize that the true cost of a lot of the cheap stuff today is in fact very high.

We need to take a step back and recall that the purpose of the economy should be to improve human well-being, not to simply increase consumption or GDP. A 2020 study led by Julia Steinberger showed that historical advances in life expectancy were driven by satisfying instrumental needs, like food and household electricity; economic growth on its own was not sufficient. Rather, it is education, and functioning social structures to meet needs like healthcare, that drives progress in human development. For industrialized countries, growth in GDP does not necessarily lead to increases in well-being.

Put simply: To have a functioning economy, we need a stable climate. Business as usual cannot continue, because it is leading to climate ruin.

Incentives in today's economic system are incredibly biased toward short-term profits, which is driving us toward long-term collapse. Profitable investments should be those that are actually good for people and planet over the long run. But standard neoclassical economics privileges instant gratification and severely discounts the value of the future. This kind of calculation favors burning all the fossil fuels we can now and making cleanup someone else's problem later (namely, future generations).

Traditional cost-benefit analysis led to the unwise conclusion that the "optimal" level of warming is 3.5°C or more. This implies it's preferable to pay to adapt instead of preventing the harm of catastrophic warming in the first place. But an ounce of prevention is worth a pound of cure, particularly when pollute-now, pay-later thinking fails to account for the facts that some losses are irreversible and irreplaceable and that climate change is rewriting the rules of the economy.

Standard assumptions about continual economic growth are not valid under climate breakdown. A study by Frances Moore and Delavane Diaz showed that most studies of the cost of climate change don't account for how much warming slows economic growth. While projections of the very long-term economic impacts of anything are necessarily speculative, Marshall Burke from Stanford led a study estimating a 23 percent loss of global GDP under high emissions. (For comparison, the International Monetary Fund projected the economic hammering from the coronavirus pandemic would lead to a 4.4 percent loss of global GDP in 2020.)

By applying a Regeneration lens to the economy, we can start to support the kinds of growth that really matter—the growth of the well-being of people and nature. Such growth may (or may not) align with growth in gross domestic product. As "growth agnostic" economist Kate Raworth writes, "Today we have economies that need to grow, whether or not they make us thrive; what we need are economies that make us thrive, whether or not they grow." You can certainly have thriving life on earth without a functioning global economy, but the other way around doesn't work. Therefore, the economy must work with the natural limits of the biosphere.

Such an economic system would provide fair distribution and efficient allocation of resources at a sustainable scale. To achieve this, we're going to need a massive shift in how we value things economically, and what we value. We need to start measuring (and designing the economy to promote) what really matters to the well-being of people and the planet. This includes recognizing how precious the tiny remainder of our carbon budget is for a good life on Earth.

The Amorality of Markets

The Economist (whose rigorous reporting I generally enjoy, as an annual Christmas gift from my dad) stated that a carbon price sufficiently high to meet the Paris Agreement would "asphyxiate" the economy. The implication was that we will just have to live with the climate consequences of inaction.

If it's too expensive to save human lives and livelihoods and the living world, then we are measuring value ass-backward.

A key shackle of the Exploitation Mindset that we need to break is letting the unfettered market dictate our values to us. Blindly following the invisible hand of the market as a moral compass leads to the brink of ruin. Neoclassical economics is inherently amoral; not immoral (doing the wrong thing), but simply lacking value orientation. Markets are efficient systems for distributing resources based on the information they have, but markets have no heart or brain; they just maximize whatever value they trade, given the rules of the game they're given.

Do you want to live on a planet with extensive, preventable suffering and death? Cost-benefit analysis is the wrong tool to answer that question; you need a moral compass instead. Market rules should set limits that protect what matters most, like human rights and the right to a stable climate and a living planet. These are the preconditions for well-being, not fringe benefits to enjoy if cost-benefit analysis decrees them profitable under today's vast undervaluation of the true cost of carbon pollution. People in power need to set economic rules fairly, guided by values that make clear what's acceptable, which should be transparently debated in society. These rules should draw from the Regeneration Mindset: Protect people and nature; reduce harm; increase resilience.

Another fundamental mistake of the Exploitation Mindset is to think that everything can and should have a price tag. We need to remember that money is a means to an end, not the ultimate arbiter of what really matters. Money is a useful human invention to make efficient trades and comparisons, but not everything should have a price tag or can be justly exchanged. Further, some losses are irreplaceable, incalculable, and impossible to compensate for—including sacred burial sites, national treasures, and whole island nations. Today's fossil dependence can be measured by its cost in lives: A 2019 study found that fossil fuels cause the majority of deaths from outdoor air pollution, via heart, lung, and other diseases. The authors conclude a fossil phaseout could save 3.6 million lives per year. There's something with real value.

True Costs of Fossil Fuels

Here's a no-brainer: Economic incentives should be aligned with rather than against a stable climate. We need policy change so that market prices reflect true climate costs, and regulation so that the root causes of climate breakdown are quickly phased out.

One policy that's currently out of whack with climate reality is heavy public subsidies of fossil fuels. Similar to the counterproductive agricultural subsidies we saw in the last chapter, taxpayers are currently spending huge sums making a dangerous product artificially cheaper, encouraging its increased use, and causing increased climate harm. Governments globally spent almost $500 billion in 2019 to make fossil fuels cheaper to produce and consume. When you include the indirect costs to health, local air pollution, and environmental damage, the International Monetary Fund estimates fossil subsidies are ten times higher, totaling 6.5

percent of global GDP. That's more of the global economy than Japan.

Proponents argue these handouts help poor households afford daily necessities, but other policies can support decent living standards for all without supporting climate destruction. There is a broad international coalition working to end fossil subsidies, while recognizing that policy reforms need to make essential energy services clean and affordable for everyone.

While fossil subsidies are most heavily concentrated in a few countries, practically every country is catering to fossil interests. Even in Sweden (Sweden! With its ambition to be "one of the first fossil-free welfare countries in the world" and its relatively clean energy system running on 54 percent renewable energy!), three times more public money financed climate-harming subsidies in 2017 than went toward the entire national environmental budget. And that was during a parliamentary session with a record high in environmental spending, with the Green Party in power.

Money going to climate destruction is money not available to invest in clean alternatives. Direct global fossil fuel subsidies were three and a half times higher than those for renewable power in 2017; society spent nearly twenty times more on fossils than renewables, when you also include the air pollution and climate costs of fossils. Despite this uneven playing field, the cost of renewables has plummeted over the last decade. A 2020 analysis by the International Renewable Energy Agency found it's increasingly cheaper to build new solar and wind projects than to build new fossil energy, or even to keep operating existing coal-fired power plants. In fact, they conclude that replacing the costliest quarter of global coal power with solar and wind would cut system costs by up to $23 billion per year.

Price or Prohibit Pollution?

Despite my criticism of the way the economy works today, I believe the market can and should play a role in helping to fix climate change, if it's mobilized fast and effectively. However, it's important to recognize that market approaches are one of many necessary tools; on their own, they will be nowhere near enough to prevent climate breakdown in time.

Given the decades of political failure to act on climate, we've simply run out of time for market approaches alone to do the necessary job. In a 2020 *Nature* article, scientists write: "In 2010, the world thought it had 30 years to halve global emissions of greenhouse gases. Today, we know that this must happen in ten years . . . for 1.5°C (25 years for 2°C)." Basically, the world now must do four times the work—or do the job three times faster. If a global carbon price had been established even as late as 2010, when we "only" had to reduce emissions 2 percent per year to meet the 2°C goal, perhaps we could have relied on a gradual increase in carbon price to sort out the climate problem for us. But now, staring down the barrel of just a few years of remaining carbon budget, that incrementalism is far too little, too late.

The High-Level Commission on Carbon Prices estimates an economy-wide carbon price to meet the 2°C goal should be at least $50–100/ton by 2030, and should already be at least $40–80/ton today. In reality, though, just one-fifth of global carbon emissions fall under any kind of current or planned carbon pricing or trading scheme, with an average price under such schemes of just $8/ton in 2018. Only 10 percent of emissions are priced at a level consistent with even the 2°C goal.

Carbon prices to date have generally been set very low, and they've made at best a marginal dent in slowing emission increases. The largest carbon market is the cap-and-trade system in the European Union, where too many permits to pollute were historically handed out (thanks to industry lobbying), making it far too cheap to keep polluting. Sweden has the world's highest carbon price, at about $120/ton in 2020. Since Sweden's carbon price was introduced thirty years ago, along with additional climate policies, emissions have declined about 1 percent per year. To meet the 1.5°C goal, the world now needs to reduce emissions *almost eight times faster*—7.6 percent each year until 2030. A carbon price isn't enough.

In the simulated climate negotiations exercise from Climate Interactive that I run with my students, they're given a two-page briefing on their negotiation positions, playing roles like "World Governments" or "Climate Justice Hawks"; then the clock starts ticking. This role-play has given me a preview of a likely outcome for carbon pricing that worries me.

In the real world, Republican leaders are increasingly advocating for a carbon price, with former Reagan secretary of state George Shultz and Ted Halstead arguing in January 2020: "The newfound Republican climate position can be summarized as follows: The climate problem is real, . . . and the GOP needs a proactive climate solution of its own." Oil companies are shifting strategy from directly blocking climate policy to publicly voicing support for a carbon tax (they're less vocal about the strings attached).

In one recent simulation, my students representing fossil companies made a carbon price their main climate strategy, in exchange for lax government regulation and other vast concessions that were good for their shareholders but a disaster for the climate. The fight over carbon pricing sapped a lot of energy and attention in this exercise, strongly polarizing interest groups and fracturing political will. After multiple rounds of negotiation, simulating years passing in the real world, the eventual global carbon price of $58/ton that my students fought tooth and nail to win was modeled to reduce warming by only 0.3°C.

Experience in the real world shows that a carbon price is not a panacea. As Bill McKibben wrote, there are no silver bullets for climate, only silver buckshot. Pricing carbon could help reduce emissions as part of well-designed climate policy packages, but it will be nowhere near enough on its own. Where it's used, carbon pricing requires careful policy design to gain public acceptance; economists argue that redistributing revenue from a carbon price as a dividend can help get citizens on board.

The Climate Doesn't Care What Carbon Costs

The science is clear that *any* ongoing carbon emissions will continue to drive climate destabilization for millennia. We need to keep in mind that the correct target for atmospheric carbon pollution is *zero*, not "less carbon pollution that someone paid more to emit"—the atmosphere doesn't care. Thus we need to recognize a carbon price for what it is: one tool among the many needed to switch from polluting to clean energy sources in time to comply with the Paris temperature goals. In the near future, coal, oil, and gas will be recognized as risky liabilities instead of valuable assets. An immediate and rapidly escalating carbon price could come in handy to help prevent a perverse scramble to burn all the fossil fuels before they become worthless. But mostly, we just have to leave them in the ground.

Academics and policy experts increasingly agree: To rapidly reduce real-world emissions, we need to go beyond economic incentives. So far, regulation to require energy from renewable sources and energy efficiency standards have proven more effective than pricing carbon at reducing emissions. Ultimately, we need to restrict and finally stop extracting, selling, and burning fossil fuels. Fergus Green of the London School of Economics argues that fossil fuel bans (any government prohibition of any part of the fossil fuel supply chain, from exploration, construction, or supply to production and financing) signal the moral wrongness of fossil fuels, build an anti-fossil norm, and encourage states to persuade others to join them, as happened with the Powering Past Coal Alliance. The independent Climate Policy Council in Sweden, which is tasked by Parliament with holding the government to account on meeting the country's climate law, recommends the Swedish government set a stop date for selling fossil fuels. A near-term fossil stop date will send a clear signal to industry that a future with a stable climate is not fossil-based. It will push current leadership to accept responsibility for starting the rapid transition needed, not expect to make it someone else's problem after they retire. In the end, the correct price for carbon emissions is infinite: a ban.

Popping the Carbon Bubble

Because the global economy today is built on fossil fuels, a lot of money is tied up in assets that are essentially worthless if we care about the climate. Ongoing investments in climate destabilization have created a "carbon bubble": sunk costs in fossil infrastructure and other carbon-based assets and capital that are climate liabilities. I'm talking about pipelines still being built that should never have fuel in them, and new coal energy plants that need to shut down before their useful life is up in order to limit warming enough to meet Paris goals. We need a plan to deal with the loss of these falsely valued fossil assets, to make sure these losses do not harm the most vulnerable. The sooner we start tackling this problem, the better our options for dealing with it fairly.

It's disheartening that many of the industries prioritized for financial bailouts due to the coronavirus pandemic, like airlines, are precisely those that need to be decreased for climate stability. Society needs a plan to shut down the polluting sources that the climate cannot afford and replace them with sustainable alternatives. Wiser public investments would promote fair transitions for workers in these industries, while scaling up jobs in green sectors, like building carbon-free transport and energy infrastructure.

Even under today's flawed market signals, though, a seismic change is already afoot. Banks, insurance companies, and investment funds are starting to refuse to fund fossil enterprises like coal mines, oil sands, and fracking, citing both financial risk to their investments, as well as climate concern.

Society will need to cough up some cash up front to invest in a stable climate, but a habitable planet is a great investment. It's a much better deal than the high price we're already paying for climate change, in everything from disaster cleanup to lost education, wages, and lives. As Yale scientist Jennifer Marlon writes, these costs will continue to rise sharply as long as we keep burning fossil fuels. A study led by David McCollum found that, to meet the 2°C goal, low-carbon investments need to overtake fossil ones globally no later than 2025, then keep growing. Meeting 1.5°C will require an even faster phaseout of fossil investments, and acceleration of investments in clean alternatives. The deal of the century is climate stabilization: According to the IPCC, investing 2.5 percent of global GDP through 2035 is required to meet 1.5°C (the deadlier 2°C is cheaper).

Investments and policy should support each other toward decarbonization. A 2018 study found the huge price decline for

renewables was due mostly to public and private R & D investments, with a dramatic contribution in the last decade from economies of scale from large-scale manufacturing, such as in China.
The specific design of Germany's renewable electricity policy, which
was highly effective in creating demand for renewables, helped enable these lower costs. National policies can have a global effect,
by spurring innovation and economies of scale that drive down
prices for everyone. Imagine what could happen if more countries
and industries actually started spending their resources and used
policies to support seriously tackling climate change.

Applying the principle of increasing resilience, investments
should be focused on strengthening decentralized, local community-
based energy (which climate journalist David Roberts calls "the
only way to truly vouchsafe the promise of safe, reliable electricity
in a warming century"). A more localized strategy would also provide a lot more jobs. My Vienna-climate-beer friend Charlie Wilson has shown that smaller-scale, lower-cost "tiny tech" that can
be mass-produced and distributed widely can drive faster and
fairer decarbonization through faster learning and innovation, as
well as reduced demand. The "many small instead of few big" investments he and his colleagues analyzed also create more net
employment in roles such as installing solar panels or retrofitting
homes to be highly energy efficient.

To sum up so far: To disentangle the global economy from the
dangers of the Exploitation Mindset, governments need to stop
making climate harm cheaper than stabilization; banks need to
stop funding fossil infrastructure; investors need to transition to
zero-carbon portfolios; and fossil energy companies need to either
rapidly retool to produce only clean energy, or prepare for a managed decline. These are some of the key policy goals to push for
as a citizen, through the means we'll talk about more in the next
chapter on politics.

I agree it sounds a wee bit daunting! But catastrophic climate change sounds worse.

Now that we've set our sights on the goals for the global and national economic systems, let's see how you personally can help make them happen through your private economy.

Put Your Money Where Your Mouth Is

Some of the biggest power to effect change you have as a consumer is not in how you spend your money but in where you refuse to spend it, and how you push on the financial systems you're already a part of. Here's how to start to disentangle your life from corporations that continue to support climate destruction and support institutions that put your money to work for a better world instead.

If we follow the money, we can go back to where a lot of it is kept: banks. A 2019 assessment of how the mega-banks worldwide stack up against the Paris Agreement is a dismal report card: All of China's banks get grade F, as do all but one of Canada's (TD squeaks in with a D−); the United States posts D− across the board; Europe leads the pack, but a D+ average isn't much to brag about. Basically every big bank is still funding an enormous amount of fossil fuels, including *still funding expansion* of fossil fuels, where Chase, Citi, and Bank of America lead the world in the wrong direction.

A few years ago, I learned that my old mega-bank (hi, B of A!) was funding the Dakota Access Pipeline; they also funded Arctic drilling and were among the top funders of extreme deepwater extraction and fracking. I was horrified. I closed my account and wrote the heads of the company and of my local branch a letter explaining my decision (I'm still waiting for an answer . . .), which I posted on social media to encourage friends to do the same. I switched to a local credit union (a member-based, non-profit, primarily local financial cooperative that returns income to members in the form of lower fees, as opposed to paying profits to investors).

In terms of my personal finances, I switched my investment account from another climate-destroying mega-bank, UBS, to my local credit union. When I looked at the carbon footprint of my stocks using the Fossil Free Funds calculator, I found my money was indirectly supporting tons of climate destruction. In discussion with my financial advisor, I divested from the high-carbon stocks and found low-carbon options he agreed were also good financial investments, including one called Etho Climate Leadership recommended by my friend Rahul for its rigorous screening criteria. It's a diversified fund of climate leaders in their respective industries, who must have at least 50 percent lower emissions than the industry standard. All fossil fuel companies are completely excluded. I also divested the portion I can control of my Swedish pension fund from fossil fuels, and stopped using my Chase card that offered bonuses of climate breakdown (free air miles) along with their investments in dirty energy.

What about the impact you can have at work? Companies have a lot of work to do to ensure that their finances and supply chains are moving toward zero carbon. Employees can have a big impact by pushing their company to set ambitious and rigorous climate goals, and put in practice the steps necessary to achieve them (or if you're the CEO, please go right ahead and do this today). One resource is Science Based Targets, which supports companies setting greenhouse gas emission reduction targets in line with climate science, so every company can do their fair share to deliver meeting the Paris temperature goals. Another is a comprehensive tool from Normative.io, which converts an organization's financial budget into the corresponding carbon footprint to help identify and start to reduce sources of high emissions along the value chain.

Individuals can also join an existing Fossil Free divestment campaign, or start a new one using the tools on their website, to pressure their employer to divest their holdings from fossil fuels and reinvest them in a way compatible with meeting the Paris Agreement. Bill McKibben, who helped launch the divestment movement, wrote, "One goal was to 'take away the social license' of the fossil fuel industry." It's working; a 2018 *New Yorker* article reported that "oil companies have started to flag the movement as a material risk in their securities filings." These campaigns have rapidly spread from college campuses to institutions ranging from the world's biggest fund manager, BlackRock, which recently announced plans to divest from coal, to the World Council of Churches. I've long supported the campaign to get my employer, Lund University, to divest, which has partly succeeded.

What about going beyond not doing harm with your money, to actively do good? I aim to donate 10 percent of my income to support a range of charities, with a focus on climate stabilization, empowering women and girls, fighting poverty, and protecting nature in ways that also protect the livelihoods of local people.

I see these donations as a way of increasing my positive impact for causes I care about, not an "offset" for the harm I've caused by the climate pollution I haven't eliminated yet, for which I take responsibility. Offsets are schemes where you continue polluting but pay someone else to pollute less. I find this approach both morally and practically problematic.

Morally, I don't like the idea of overconsuming more than my share while paying someone else to go on a carbon diet for me. Further, many offset projects are imposed in ways that don't truly benefit local communities.

Practically, an EU-commissioned report found only 2 percent of the offset projects assessed had a high likelihood of delivering real emissions reductions that would not have happened otherwise, compared with 85 percent with a low likelihood to do so. They recommended supporting results-based climate finance instead.

After I've reduced my own emissions as much as I can, one option I like for dealing with my remaining emissions is a "future box." Every time you spend money on fossil fuels (fill up the car, buy a plane ticket), put an equivalent amount of money aside for your future box. This is money to invest in future reductions of *your own emissions* toward zero (like replacing a propane heater with an electric heat pump). Alternatively, you could give this money to worthy causes that are working for broader social change, without trying to justify the benefits they provide to people and nature to excuse the tons you emitted. Given the vast undervaluation of the true cost of carbon today, offsets are usually too cheap. Doubling the cost of gas and flights is much closer to the true social price, meaning you will spend more money toward a worthy cause than you would through an offset, and you're also more likely to reconsider what is truly necessary carbon spending.

Manage What We Can't Avoid; Avoid What We Can't Manage

As we retool our economies to align with regeneration, we'll need to make investments that reduce harm at the source (get to zero emissions fast, as well as reduce non-climate risks) while we also strengthen resilience (prepare for worsening climate impacts and plan to protect the vulnerable). To live in our warming world, we need to avoid the changes we cannot manage, and manage the changes we cannot avoid.

What does this mean for a climate-changed California? In a 2014 study, wildfire scientists advocated reducing fire risks, including by removing fuel through prescribed burns, but argued for a paradigm shift away from the impossible goal of complete fire suppression and control and toward living better with wildfire. Homeowners reducing the ignition potential of their home (for example, through making buildings more fire-resilient and defensible, and removing ignition sources like gutter debris) is the most important action to prevent fire from destroying homes. Under increasingly common extreme conditions where fire behavior overwhelms available firefighters and resources and the spread of fire cannot be prevented, the idea is to make the community itself the firebreak.

After hard experience, California is better preparing its emergency fire response, but the overall fire strategy is still desperately behind what's needed to prevent massive suffering. The year after my hometown burned, devastating fire returned to California. A few hours north of Sonoma, the Camp Fire burned the town of Paradise to the ground in the deadliest and most destructive fire in California history. The insurance firm Munich Re reported the Camp Fire as the most expensive natural disaster in the world in 2018. Then 2019 ushered in more losses from extreme weather and climate events, when another set of California fires were one of fourteen catastrophes in the United States with economic losses over $1 billion each.

One of the 2019 fires, the Kincade Fire, again set wine country ablaze. Numbly reading the news, I saw a photo of a woman, who looked like a suburban soccer mom, walking down a highway in flip-flops, pushing a wheelbarrow filled with her remaining earthly possessions. My friend Mara wrote on Facebook, "Climate refugees fleeing just 50 miles from San Francisco." That could have been anyone I love. That could have been me.

Fifteen percent of Californians live among the dry, remote hills and forests in areas classed as extreme fire danger. Some analysts are already saying these areas should be abandoned for human habitation because they're too expensive and impractical to protect under California's new climate-charged fire regime. Journalist David Roberts wrote, "Denser, safer areas of the state will have to make room for them." Meanwhile, the governor opposed limiting housing expansion in remote wild areas that spread people out among tightly packed dead trees, claiming it was against California's "wild and pioneering spirit." No politician wants to fight the real estate lobby.

But the climate doesn't listen to lobbyists, and the sheer economics of increasing climate costs are starting to speak for themselves. Homeowners insurance companies in California paid out $1.70 for every dollar of premiums they collected in 2018 and are balking at insuring an increasing number of seaside and floodplain homes in the United States. As climate change makes insurance more necessary, it is rewriting the bottom line for insurance companies. Their old business model is breaking down.

I was unceremoniously dumped by my insurance company in August 2020. I inherited a farmhouse on my grandparents' old turkey ranch, where I lived for five years and have imagined retiring. Precisely as a new gang of catastrophic wildfires were tearing through wine country, raining ash and turning the sun an apocalyptic orange through the lung-searing smoke, I opened a letter: YOUR PROPERTY IS LOCATED IN AN AREA SUBJECT TO HIGH RISK FOR WILDFIRE, WHICH INCREASES THE RISK OF FIRE OR SMOKE DAMAGE. AS A RESULT, THIS POLICY WILL BE NON-RENEWED EFFECTIVE 12/6/2020. I felt a cold lead weight in the pit of my stomach. How could I plan to grow old in a house too dangerous to insure? How could it be safe for anyone to live there?

The financial case for climate action is overwhelmingly clear, and the costs of inaction are intolerable. Beyond dollars and cents, homes and livelihoods and ways of life, even whole cities and countries are at risk from climate loss. These are prices that are too high to pay. We simply can't afford not to fix it. We just can't expect that climate change will be solved through economic tools alone. We have to retool our economy to protect what we really value most, by keeping fossil fuels in the ground while increasing resilience to face the risks already elevated from warming.

Chapter 12

The Personal Is Political

Citizen Action Can Effect Global Change

In Paris in 2015 for the last day of the annual United Nations climate negotiations, I managed my expectations as I dropped my business card in the hat. There were several times more observers here than there were tickets available for us to witness the upcoming session, which I had heard could last until the next morning. I had my phone charger and toothbrush in my bag; I was ready. The closing words from former French prime minister Laurent Fabius at that morning's session still rang in my ears, as much for their content as for the tremble in the interpreter's voice in my headphones that marked their stakes: "The world is holding its breath. It counts on all of us."

My name being called interrupted my reverie. I leaped out of my seat to collect my ticket. I arrived early and took a front-row seat in the bleachers for observers at the back of the cavernous former airplane hangar where negotiations were being held. The energy was tense and excited. Rumors were flying around the room and on Twitter: Ban Ki-moon was landing in a helicopter; a misplaced "shall" (which would require approval by an unfriendly US Senate) instead of the softer "should" was threatening the whole deal (that one was true!). Then negotiators and heads of state started coming in, hugging and taking selfies; the mood was jubilant. After decades of setbacks, delay, and disagreement, after the last two sleepless, frenetic weeks of negotiation here in Paris, did we finally have a climate agreement that the countries of the world would accept?

We did. On that last night in Paris, governments held to account by civil society achieved a more ambitious agreement than many had thought possible even days before. The agreement gave the world its mandate of political will (and gave me a great excuse to hug the incredible Christiana Figueres, who brokered the Paris deal with determination and empathy). In the jubilant celebration that followed the agreement's adoption, my fellow attendees and I danced on tables until six A.M.; climate victories deserve a proper celebration! I somehow left behind a great black cocktail dress and my favorite pair of rainbow cat leggings in my bleary rush to the train station.

The moment of international climate solidarity I witnessed in Paris was a defining personal milestone for me. But to ensure that the goals set there become reality, it's up to us as ordinary citizens to take the collaborative spirit, and the ambitious promises, of Paris to our own city halls, to Congress, and to the streets, to hold governments to account. We need to harness the power of politics to enable, support, and incentivize all the changes we've talked about so far to make decarbonized, regenerative energy and food systems a reality at a collective, societal level. Aligning politics with climate stabilization asks us to become what sustainability scientist and three-term Minnesota state representative Kate Knuth calls climate citizens, which includes "putting climate change in the center of engagement with public life."

Politics plays out through a combination of top-down leadership and grassroots civil society pressure, from decisions made in boardrooms and at kitchen tables. Politics is classically defined as "who gets what, when, and how," but it also encompasses how people battle for influence, exercise power, make decisions, resolve conflict in groups, and define what is worthy of public attention, debate, and resources. Ultimately, politics encompasses the interaction between what society demands and governments deliver. To direct our political energy toward a stable climate, we must first understand the system of carbon lock-in we need to overcome, then use the principles of regeneration to guide climate action and set political goals. Then we need policies and processes to realize these goals at scale.

Overcoming Carbon Lock-In

Why haven't we cut emissions more already? Karen Seto and colleagues write that today's high-carbon system is not purely chance; actors with social, economic, and political power actively work to reinforce the status quo that favors them. Further, interactions between today's high-carbon infrastructure, policies, and lifestyles and culture tend to entrench further carbon lock-in.

Fortunately, the politics of how to break this vicious circle are becoming clear. Steven Bernstein and Matthew Hoffmann argue it will take three political processes. First, cultural change, where new mindsets and practices shift expectations and norms; second, building capacity to decarbonize, through everything from education to institutional reforms; and third, using incentives and social movements to build coalitions that bring together and empower the many but diffuse interests who would benefit from decarbonization.

More specifically, breaking carbon lock-in will require contesting the entrenched financial and political power of fossil energy companies, the sector that topped tech to lead the global revenue list in 2019, and includes five of the ten wealthiest companies on Earth (numbers seven and ten are car manufacturers, which run on their products). These companies have been putting their money to use to block, delay, or weaken climate policy around the world, to the tune of $200 million each year for just the top five oil companies, according to a 2019 InfluenceMap report. In the United States, a study by Robert Brulle showed that lobbyists from fossil fuel producers, transportation, and utilities spent ten times more money on climate lobbying from 2000 to 2016 than environmental and renewable energy groups. Unsurprisingly, he concludes that lobbying expenditures at a given time "appear to be related to the introduction and probability of passage of significant climate legislation." A 2019 study estimated that lobbying against a single climate policy helped prevent its passage, costing the United States $60 billion in foregone social benefits. The Stanford climate history researcher Benjamin Franta puts it bluntly: "We used to think passing climate policy was just hard. But now we know there's a highly funded, well-organized campaign by the fossil fuel companies to kill these policies."

Beyond Political Feasibility

Science tells us what is technically necessary in the race to stabilize the climate; but what is politically possible? My mom (a retired politician) likes the saying that politics is the art of the possible. In that case, our challenge in this decade is to rewrite what is considered politically possible, because we cannot rewrite the laws of nature. It won't be easy, but it can and must be done.

We know that the path we've taken to the present, all the choices made by business as usual under the Exploitation Mindset, is headed for catastrophic warming, so we should not be surprised that extrapolating historical trends does not paint a rosy picture for our climate future. My colleagues Jessica Jewell and Aleh Cherp find little evidence from history that the world will globally undertake the portfolio of measures necessary to make 1.5°C. They find some evidence that rapid scale-up of renewables, phase-out of coal power, and diffusion of existing tech like electric vehicles at sufficient speed and scale might be feasible. But they call the required reductions in energy demand and improvements in efficiency "historically unprecedented." In short, they doubt the recipe for feasibility is met: enough actors capable of bearing the costs of decarbonization under specific national and international circumstances.

However, they also recognize the limits of looking to the past to do something unprecedented, writing: "Most evidence of political feasibility of climate action is derived from historic or present experience. However, if a certain solution [has] not occurred in the past this does not necessarily mean that it is not politically feasible in the future. . . . In contrast to most 'hard' constraints common in natural sciences and engineering, political constraints are 'soft.'"

If we can't look to the past for a guide to a future we want, what can we do instead to set the social goals that politics must deliver?

One approach to setting political vision is participatory: get diverse stakeholders together to brainstorm and agree on desired visions based on their values. Researchers in Sweden recently used this approach to develop a vision for sustainable food production and consumption that's now being used to inform national food policy. Visions can also be developed through art, as my former student Emma Johansson has done. Her participatory paintings made in collaboration with local artists and villagers in

Tanzania visualize the past, present, and desired future of their village and its surrounding mountains. The process of creating something together catalyzes energy to make the vision reality. These paintings evoke emotions, clarify relationships, and communicate in ways that a graph never would.

Once a vision is agreed on, backcasting can help make it a reality: starting from where we actually want to end up, then working backward to find the pathways from where we are now to where we want to be. A Swedish research group used this approach to identify a range of pathways that could keep emissions in line with 1.5°C while shrinking the footprint of land use and maintaining good social welfare and democratic political influence. They identified different possibilities from automation to local self-sufficiency to collaborative and sharing economies, sparking political creativity in starting to move toward these visions.

Visions are important, but they have to be implemented in order to stabilize the climate. The political changes we need to make this happen are heroic, so there's no time to waste. Let's look at what we know about what's needed and what works to change policy and laws.

Elect and Talk to Your Rep

According to a study where members of Parliament from eight European countries ranked the effectiveness of different forms of political participation, the top choice was to vote. The lawmakers in power matter to the laws that get passed. And we need *way* bolder politicians; a July 2020 analysis showed the last two decades of climate laws had avoided (drumroll, please) just one year's worth of climate pollution (*womp-womp*). Use your vote to remove politicians beholden to carbon lock-in (see rankings from NGOs like OpenSecrets.org) and elect those with a strong climate agenda, who have taken the No Fossil Fuel Money Pledge. Once climate-enlightened politicians are in power, they have a chance to regulate the industries driving the problem and lobbying against its timely solution. You can see the environmental and climate voting record of politicians from the League of Conservation Voters National Environmental Scorecard; a study found US states electing politicians with good LCV scores significantly reduced their carbon emissions. And while you're at it, please vote more women into office; a study led by Astghik Mavisakalyan in 2019 found that electing more women caused stronger climate policies to be adopted, resulting in lower national carbon emissions.

Please don't stop at casting your ballot, though. Politics is a constant work in progress, so it's important to remember how fragile even this top-ranked strategy is, when new administrations can undo climate progress. It's not just nature but also human institutions like democracy that need to be continuously regenerated. For example, we need to be vigilant in unfalteringly upholding and expanding free and fair elections with universal enfranchisement.

The next most effective action to influence politicians is to get active in a political party (or climate organization). Check out local chapters of climate movements like 350.org, Sunrise Movement, Zero Hour, Fridays for Future, and more to find one you like; look for homegrown initiatives nearby; or start this conversation with your neighbors. As we'll see below, a vigorous civil society helps create political will and embolden politicians to be true climate leaders.

The members of Parliament also rated creating media attention, writing letters or e-mails to politicians, and political demonstrations as highly effective, followed by petitions (which were viewed as much more effective by citizens than by politicians), internet discussions, and boycotts. (The study didn't ask about calling legislators, but former US congressional staffers report that phone calls to the local district office of your rep, focused on your personal story related to the issue, are even more effective than writing.)

Even though constituents directly contacting their elected representatives is a highly effective political strategy, it's sorely lacking today. Research by Rebecca Willis from Lancaster University found that UK politicians understood the need for climate action but felt very little pressure from their constituents to make it happen. Politicians need a broad-based and loud group of constituents demanding that they take urgent climate action, to make the issue more mainstream and to reassure them they won't lose their jobs for supporting bold climate policies. And constituents who lead by example in demonstrating a fulfilling low-carbon life reassure politicians their policies will work and be accepted. My friend Lucy rallied her neighbors with some coaching from advocacy group The Climate Mobilization, and through calls and letters convinced her congressperson to cosponsor a national declaration of climate emergency—gold climate star for Lucy!

Political Demands

What should concerned climate citizens be asking their politicians for, and what words from a political or business leader should set off alarm bells in your head, warning that they're not serious about meeting the climate challenge?

What growing groups of citizens need to demand from their leaders at every level is a plan for how they will do their share to stop warming below the Paris temperature goals, including eliminating fossil fuel production and consumption as soon as possible.

Following Regeneration principles, this will address the major cause of climate breakdown at its source; it must also draw from the remaining principles of centering people and nature while increasing resilience. High emitters should recognize that effective policies are going to require changes from us; the fairest policies will protect the needs of the vulnerable.

As we've seen, citizens should be wary of pollute-now, pay-later proposals. Any plans that dodge the simple, straightforward truth about what we're going to need to do as a species to stop climate change—stop burning fossil fuels—aren't serious. Remember, climate stabilization demands *absolute* emissions reductions (reduce the tons of greenhouse gas emissions tomorrow compared with today's baseline). The cornerstone of any corporation's or government's climate plan should be how they will reduce their emissions to zero. Their plan should demonstrate how they'll get nearly all the way to zero emissions before they start emphasizing aims to be "carbon neutral" or use offsets, mention vague "negative emission technologies," or rave about tree planting. Don't get distracted by these nice-sounding efforts if they're used to wave away ongoing climate pollution.

Be prepared to counter with demands that will actually work to stabilize the climate. Climate professor Kevin Anderson has a three-phase strategy to deliver Paris commitments in wealthy countries that sums up a lot of what we've covered so far: First, immediately reduce carbon overconsumption from high emitters through fast-acting behavior change, focusing on the high-impact climate actions (car-, flight-, and meat-free). Second, in the near term, make everything meet extremely high energy-efficiency standards through stringent emissions standards for cars, power stations, new houses, and buildings. Third, in the medium term, undertake massive construction of a zero-carbon energy supply, and electrify everything to run on it. Kevin suggests phasing out fossil fuel extraction and banning its advertisement and the expansion of fossil-based infrastructure, including airports and roads. Meanwhile, incentives should be changed to make high-emissions things (like flights) much more expensive and near-zero emissions options (like e-bikes) cheaper. To dive deeper into the wonderful, wonky world of climate policy, see resources like David J. C. MacKay's *Sustainable Energy—Without the Hot Air* and *Designing Climate Solutions* by Hal Harvey, Robbie Orvis, and Jeffrey Rissman.

Climate policy is not only national and international; it's fundamentally local, and we need to push for local solutions as well as global ones. Local city councils, for example, are constantly considering proposals that ultimately affect the climate—from zoning to infrastructure to food policy. For example, informed climate citizens can push their city to adopt and implement policies that prioritize people over cars in the city center and help municipalities prepare for flooding, fire, and other climate-related hazards. Local- and city-level efforts, like the grassroots Transition Towns and the bigger C40 Cities Network, are critical for creative climate experimentation, learning, and leadership.

Fairness and Just Transitions

Great, you've gotten your politicians on board with the critical need for climate policy! Now: What kind of policy will work in practice?

It sounds like going back to kindergarten, but social science research shows that a critical ingredient of successful policies is that the actions they demand and the benefits they provide are perceived as fair. The public needs to understand the policy consequences and alternatives and believe that both the distribution of consequences and the political process used to make the policy are fair. Since fairness requires social buy-in, we need to be having social debates about the fairness issues at stake, and we need leaders who explain the urgency of action, balanced with the need for inclusive deliberation.

To be successful and legitimate, climate action must be something that happens *with* people and *by* people, not *to* people. Geographer Karen O'Brien writes that climate action needs to recognize people as agents of change, rather than treating them as objects to be changed. Creating and enacting effective local climate policies requires the participation of the individuals affected, so they can contribute to developing and modifying the rules and will perceive them as fair. As climate justice advocates say, "Nothing about us, without us." Alongside centering human interests, remembering from Regeneration to include and care for nature might take the form of involving scientific and traditional ecological knowledge in the decision process, or even granting rights of personhood and legal standing to natural entities, as New Zealand gave the Whanganui River in 2017.

One promising strategy to make more fair decisions is citizens' assemblies, like Climate Assembly UK, a group of citizens selected to represent the UK population in terms of demographic factors and their climate attitudes. The members met over five months for briefings from experts and facilitated, in-depth discussion among themselves, then wrote recommendations for what the United Kingdom should do to meet its legally binding climate target of net zero by 2050. Their recommendations will be reviewed by UK parliamentary committees and debated in Parliament. Another good example of inclusive institutional innovation comes from Wales, where the world's first future generations minister is working to ensure that the interests of citizens not yet born are represented when political decisions are being made.

An important aspect of fairness is a just transition for workers whose current jobs endanger the climate for everyone. Voters care about jobs; a 2020 study in the United States showed that including employment provisions like a job guarantee, retraining fossil fuel workers, and unionized clean-energy jobs strongly increased support for climate policy among Democrats and slightly among Republicans. Fossil industries actually provide surprisingly few jobs; for example, Arby's restaurants employ more Americans than the entire coal industry. Just transition initiatives are already underway; for example, Spain has a Ministry of Ecological Transition that offers job retraining or early retirement for former coal miners, and workers from Alberta's oil sands are mobilizing for retraining in clean energy jobs.

The Court of Public Opinion and the Courtroom

A free and independent press is critical in many ways, including shaping the court of public opinion. A study of US public concern for climate change found more media coverage directly increased public concern. The strongest effect was congressional attention on climate change, which caused greater public concern through increased media coverage. But media often miss opportunities to explain the climate links in their stories, and especially to highlight solutions. A positive example comes from the daily Swedish newspaper *Sydsvenskan,* which told every story in every section from a climate angle in an August 2018 edition.

Media also need to look in the mirror and examine the role they're currently playing in supporting climate breakdown. When media continue to feature ads for high-emitting companies, it both promotes climate harm directly and also undermines the credibility of their reporting on the climate crisis. For example, in their September 21, 2019, climate issue, *The Economist* called for climate action and called out climate misinformation, writing in two separate stories, "To reduce its climate risks, the world needs to curtail its production of oil" and "doubt about greenhouse warming, which the fossil fuel-lobby deliberately fostered." Yet halfway between these stories, they featured a full-page advertisement from ExxonMobil, which we saw in Chapter 7 led public misinformation campaigns and invests a pittance in non-fossil energy, but a substantial sum falsely advertising itself as a climate leader.

Other newspapers are showing leadership in refusing to give their platform to climate polluters. The Swedish paper *Dagens ETC* stopped accepting fossil fuel, car, and aviation ads in September 2019. *The Guardian* followed partway in February 2020 by banning fossil fuel ads. Signals like these help create the norm shifts that increasingly view fossil fuel companies as the tobacco companies of the 2020s, with car and airline firms not far behind.

Legal cases are also playing a growing role in driving climate policy. More than thirteen hundred climate-related lawsuits have been filed worldwide as of 2019, most in the United States and most targeting governments. However, high-emitting companies are increasingly targets, including for failing to account for and disclose climate risk in their business plans. In the Netherlands, the nonprofit group Urgenda won a 2019 lawsuit arguing that the government had a human rights duty to protect its citizens from climate change. The ruling mandated emissions cuts of 25 percent by the end of 2020, compared with 1990 levels, and national policies in line with the 1.5°C goal. The ruling is likely to close coal-fired power plants ahead of schedule.

Laws and civil society work together to create social change. Lawyer Sophie Marjanac argued at the 2017 UN climate conference that campaigns in the streets to change hearts and minds create the conditions so that judges or a randomly selected group of jurors are likely to support change. She sees the laws that result as buttressing rather than leading social change, with new laws quickly becoming normalized and accepted, and a new generation growing up with them as the new normal. Thus, climate wins require social movements from civil society.

The Moral Compass of Civil Society

To get politicians to act, civil society needs to make nonviolent, urgent demands that are impossible to ignore. As Barack Obama said, following a global wave of People's Climate Marches in September 2014, which included a reported four hundred thousand people in New York: "Our citizens keep marching . . . we cannot pretend we do not hear them." Two months later, he announced the bilateral US-China climate deal. This deal was important in paving the way to the Paris Agreement, signaling to the world that the United States was on board after a long history of being one of the main blockers of international climate action. (Sadly, we reclaimed this position when Trump left the Paris Agreement, against the wishes of almost 70 percent of Americans, but thankfully Biden pledged to rejoin on his first day in office.)

As part of the September 2014 mobilizations, my students organized a climate protest by bike in Lund. Until just before, I had not planned to attend. "I'm a scientist, not an activist," I thought to myself. I had never been to a protest or marched in the streets for anything, and it sounded a bit . . . weird? Silly? Vulnerable?

Then I read an article that changed my mind. (I've looked for it and can't find it, so that author will never know they inspired me!) Basically, it said, what if *this* was the thing that made the difference? There's no way to know beforehand, and you'll probably never know afterward either, but what if this is the start of the tipping point? Wouldn't you want to be a part of that? "Okay," I thought. "I'll go."

I sat at my kitchen table with a piece of poster board and colorful markers and pondered what to put on my sign. I wasn't sure what my political demands should be, so I ended up making a footnoted protest sign summarizing the science, using the framework I've now taught for more than a decade: It's warming. It's us. We're sure. It's bad. We can fix it. I added footnotes for references; claims need to be supported by evidence!

At that first protest, we experimented and learned what worked (signs, slogans, catchy chants). Emboldened, many of us joined a bigger protest the next day in Copenhagen, filled with signs that said #WorthSaving, listing all the things people cared about that were at risk from climate change. One of my students, Joep, seized the moment and the microphone as the assembly was winding down, and urged the crowd to follow him through the streets so the message would reach more people. We followed, including another of my students who had just fallen in love with him for his courage (last I heard, they're still together). A protest can be a wonderful place for romance.

It gives me chills of hope to see how these small gatherings, which at first felt like a hardy band of friends howling into the abyss, have quickly become crowds in the millions worldwide, as youth movements like Zero Hour and Fridays for Future galvanize a new generation. The youth-led school strikes for climate got more than 6 million people worldwide out on the streets in September 2019, an exponential rise from a few years ago. Years of patient local organizing and coalition building have laid the groundwork to make possible today's rapidly increasing climate youth movement, which I've supported by participating in demonstrations, giving speeches, and writing op-eds.

I've been inspired to see how just a few people can start something that ripples out to become much bigger. One of the most striking examples was hearing Bill McKibben describe the humble beginnings of the NGO 350.org, which has helped energize climate protests worldwide. He was sitting around a table with seven of his Middlebury College students in 2008, saying, "We have to do something about climate change." They brainstormed for a while and decided to focus on a worldwide day of climate demonstrations. Each student took charge of organizing actions on one continent. The student in charge of organizing Antarctica also got assigned "the internet" for good measure. As Bill McKibben sees it, the most important thing someone can do as an individual is "not be an individual . . . we need to join with each other in movements."

Citizen involvement is critical to driving institutional, private sector, and political change. After protests and petitions from students and faculty, leaders from universities across Sweden adopted a climate framework outlining thirteen priority areas and pledging to implement measures to be in line with 1.5°C by 2030. There's a long way to go to implement the necessary policies in practice, but at least they are on the table. Similar plans exist or are being developed for many industries; every industry needs one.

In Paris, I saw firsthand how creative and passionate civil society protests at the negotiations made a critical difference for getting the 1.5°C goal in the final text. One Dutch negotiator I spoke with called civil society NGOs the "moral compass" behind the Paris Agreement. Inside the negotiation halls, it truly felt like the eyes of the world were watching, thanks to screens playing scenes from demonstrations that had been held worldwide in the lead-up to Paris. NGOs and public campaigns are critical in representing public interest in politics, holding governments to account, and building capacity for citizens to become effectively engaged.

Activism works. Your participation matters to social movements. When we get a critical mass to stop supporting the current system and start working toward a better one, change can come surprisingly fast. University of Pennsylvania research found it takes only 25 percent of a population to commit to a new norm in order to tip majority opinion in their favor; therefore, these first actors are disproportionately critical. Their momentum helps create widespread changes in social norms and behavior, which can catalyze rapid social change. Social movements and political campaigns have proven critical in making social change happen. There is room for everyone within these movements, and the people behind them would love to have your support.

On the Streets of Stockholm

In September 2019 I was in Stockholm during the youth climate march (with adults invited to join in solidarity). The strike had been added to the afternoon's agenda for the workshop on climate ethics I was attending. I was deeply cheered that academics were mobilizing from the ivory tower to the street, compelled by the rising drumbeat of our results and the moral clarity of young people.

I got goose bumps as soon as my colleagues and I came up from the subway station and joined the throngs. After five years of joining climate marches, bike rides, and other actions, it felt transformatively different for me to be part of such a huge crowd. Now we weren't the minority, with everyone else looking in on us. We were the whole city. The crowd was so thick it took half an hour before we even started moving. Once we were walking, more people kept coming behind us, and it was impossible to see the end of the flow of people. We passed shops proudly proclaiming CLOSED

FOR CLIMATE STRIKE. I talked to a grandmother along the way; this was her first protest ever. She simply said it was time.

Our march passed the Swedish Parliament, where Greta Thunberg first sat alone on the pavement for her School Strike for Climate in August 2018. Just over one year later, we were part of a record-breaking crowd of more than sixty thousand people, peacefully and creatively and urgently demanding our government follow the Paris Agreement. The government cannot ignore that many people for long.

Politics helps shift culture and organizes and directs collective action. We have to use the power of politics to translate the Paris Agreement goals into policies that will drastically cut carbon while they protect and improve well-being. We need climate citizens to come together in order to reach a political tipping point where we'll overcome vested interests to address the climate crisis with the urgency and fairness it demands.

Chapter 13

Being a Good Ancestor

Building the Climate Cathedral

S hortly after I moved across an ocean to Sweden to start a new job, on a continent where I had no friends or family, while on the brink of a painful divorce, my friend Eve came to visit. I was *very* glad to see her.

We decided to pay a visit to one of Eve's relatives, a man named Sture whom she'd never met. According to a sixteen-year-old letter from Sture to one of Eve's cousins, he lived in a town an hour away from my new apartment. As we boarded the train to the unfamiliar station, Eve grabbed my hand and said, "We have no idea what we'll find when we get there. This trip could end in a graveyard."

Instead, after a few wrong turns, we found ourselves standing on the doorstep of a tidy yellow house. A bit apprehensively, Eve knocked.

There was no answer. She knocked again. Still nothing.

As Eve raised her fist for a third knock, the door flew open. There stood an elderly man with abundant white hair and gold wire-rimmed glasses that would have been hipster on someone six decades younger.

He didn't seem that pleased to see us, as he was already shouting at us in Swedish and waving his hands emphatically as if we were crumbs that he was brushing off a table in front of him. We tried to greet him, but he wasn't listening to a word we said.

Eve fumbled in her bag and pulled out the letter, offering it to him while pointing at his name and address in the sender's corner, then at him. "Excuse me, is this you? Are you Sture?"

The man glanced quickly down at the letter, then back up at Eve, his tirade still going full force. Then he did an almost comical double take: a longer look at the letter, a longer look at Eve, the torrent of Swedish starting to slow down. Then an even longer look at the letter, a long searching look at Eve as his speech slowed, then stopped.

Then he threw his arms wide and proclaimed in English, "You are relatives! From America! I thought you were Jehovah's Witnesses. Come in, come in!"

Sture served us coffee and biscuits, inquired about Eve's family back in the United States, and shared the sad news that his beloved wife, Gun, had recently died of cancer, a few years after he suddenly went completely deaf. Now he lived alone, and his isolation was plain.

He said he wanted to show us something upstairs. I was expecting a family photo album, but instead he unrolled an enormous scroll, about two feet wide and more than twenty feet long. The scroll contained more than four hundred names he had painstakingly lettered across seven generations of his family tree. He told us he had pieced together the relationships through what a fellow pensioner in his English class had called his "gynecological" (she meant genealogical) research. Toward the bottom, he found what he was looking for: Eve's place in the family tree. "Ah! Eve-Lyn, you are from the tribe of Hildegarde!"

Sture became my adopted grandfather in Sweden. Until he

died five years later, we enjoyed regular visits. We would take his beloved *fika* break with coffee and cookies at precisely two thirty P.M., go for walks in the woods near his house, and pick flowers in the garden he tended meticulously. With my own grandparents long gone, I appreciated hearing his stories about growing up in a different world, but one that was clearly the progenitor of our own.

Hearing stories that spanned seven generations of Sture's family history was an antidote to today's short-termism. To meet the challenges we face, we desperately need to lift our sights beyond ourselves in the here and now, past the horizon of quarterly returns and four-year election cycles. Situating our individual stories in the wider context of the human family will help us to use our precious time wisely now and in the critical decades ahead to work toward a world we will be proud to leave behind as our lasting legacy.

The Urgency of Eternity

We are now living in the future that was hypothetical to the people more worried about the Y2K bug than the continuation of the biosphere. The failure to focus and act on climate with the urgency demanded is driving our deadlier heatwaves and fires and our rising seas.

This decade is a race between two tipping points that will flip the world into a new state: either a change in mindsets leading to a positive social tipping point supporting the transformations needed to stop climate breakdown, or a gang of looming, catastrophic ecological tipping points, system flips that could set off chain reactions beyond our control, destabilizing the biosphere and climate and destroying civilization as we know it.

We have to be the generation that acts with urgency to

safeguard the long-term future, the backdrop to the rest of our lives and to the lives of generations of people we will never know. To do so, we must simultaneously slow down and speed up. We need to slow down from the daily hustle to consider our lives against the millennia over which those of us alive today are determining the fate of the whole planet. And we need to direct our limited time with urgency and focus toward creating the world we want.

Shifting the story toward regeneration requires that we readjust our relationship with time, to align our actions to be on the scale of regenerative cycles for the planet and those living on it. We need to transcend the mindset of quick fixes to imagine the impact of our choices on the scale of the natural limits and cycles we have to learn to live within: the season it takes to grow a warm tomato that explodes with flavor fresh off the vine; the lifetime it takes to grow a mighty tree; the millennia it takes the machinations of the land, sky, and sea to circulate carbon.

Cultivating this sensitivity to time is essential to overcome the Exploitation Mindset view of time as something to be minimized on the way to maximizing production and profit, speeding past the moment in fast vehicles, eating fast food. This road leads to maximizing short-term material gains at the expense of our imperative to pass on a planet capable of sufficient regeneration for future generations.

Find Bright Spots and Embrace Failures

To act with urgency, it's too late to wait for others to act before we take the first step. As a society, despite heroic efforts from the margins, we have not even started to seriously try to stop climate change. There are lots of inspiring people fighting as hard as they can, but many more who aren't yet engaged or aren't sure what to do.

In their book about how change happens, brothers Dan and Chip Heath suggest finding bright spots, rays of hope that succeed even amid all of today's problems. Figure out what they're doing right and help make it spread. They also point out that it's impossible to perfectly script the plans along the way, since so much will change unpredictably. What's needed is a clear vision of what success would look like (zero emissions, a society that works for everyone, thriving nature); then we can use what we have available right now to start pursuing it. We have to dare to start, all of us, now, today. What we learn along the way will open many new doors.

When trying something new, we should expect and learn from failure along the way. I recently spent an afternoon with my friend Emily at a bouldering gym, attempting to climb a project several notches above our usual level. The goal was clear: the final hold at the top marked with red tape. But the last move to get there involved a huge dynamic jump that looked frankly impossible from where we sat. We discussed strategy for a minute, then just got started trying, throwing ourselves inelegantly in the general direction of the final hold. On something like my eighth attempt, I actually jumped high enough to touch it, but I was so surprised by and unprepared for the possibility of success that I didn't manage to grab it before falling.

Other climbers joined in; this growing community encouraged every attempt and inspired one another with new ideas. Through our repeated failures, Emily and I were learning, and our failure was useful to others, including a relative newbie who was thrilled to get the problem on his first attempt. Smiling, he said, "Watching you two gave me some ideas." The seemingly effortless success of more experienced climbers belied the years of previous failures they'd notched to succeed this time. Watching others try and fail, or try and succeed, inspired ever more people to join in and find what worked, in a way that sitting around all day talking about how to climb does not. Climate action is like rock climbing: We just have to get started; we'll learn as we go.

Being a Good Ancestor

To start rewriting the current story of exploitation toward regeneration, we can look for wisdom from both the past and the future. At the UN climate negotiations in Bonn in 2017, a young Maori woman told me, "The person I want to be is someone my ancestors would be proud of." She saw her own story as connected to both a past and a future filled with people she cared about. The author Bina Venkataraman suggests being a good ancestor requires imagining the future with empathy, listening to the voices of the future, and leaving heirlooms for future generations to steward.

To become good ancestors, we can and must overcome the tendency toward present bias favoring short-term gratification. Hal Hershfield, a marketing professor, believes present bias is caused by being "estranged" from and failing to empathize with our future selves, making saving feel "like a choice between spending money today or giving it to a stranger years from now." But he found he could induce people to act today more in the interest of the future. When college students saw "aged avatars" of themselves in a mirror, digitally manipulated to appear fifty years older, they chose to save more than twice as much toward retirement than those who did not.

Climate campaigns are increasingly tapping into ways to empathize with and imagine the future. For example, activists from Sunrise Movement dedicated climate time capsules in state capitals across the United States in November 2017, to be opened in fifty or a hundred years. They invited citizens to contribute "things we love and are fighting to protect" and "letters to the future"; politicians to contribute their commitments to clean energy and stopping fossil fuel development ("Those who don't will be remembered for their cowardice"); and the press to submit their coverage of climate action ("If they don't, we will leave a record of their inaction in the time capsule"). These kinds of gifts to the future hold us accountable for our actions now and force us to look beyond the next few years in office, toward our long-term impacts.

Another project that lifts our sights to the horizon is Future Library, which is collecting one unpublished manuscript each year over one hundred years. Authors to date include Margaret Atwood and Han Kang. The stories will be held in trust, unread, until they are released in an anthology in 2114. The paper on which they will be printed will come from a thousand towering Norwegian spruce trees, which were planted outside Oslo as tiny saplings in 2014. Those trees will absorb the atmosphere of the sky we make, and that our children and grandchildren make. The texture of the paper that the Future Library is printed on will reflect our choices. The Future Library is the opposite of instant gratification, or even delayed gratification; it is, rather, gratification that comes from slowness and waiting, from living today in a way that acknowledges how much the future depends on us. We who happen to be alive at this moment are writing history for so many for such a long time to come. The Future Library pushes us to write a better story.

Forget Future Generations

For guidance from the future, we don't have to imagine what hypothetical descendants would want; we can instead listen to the rising chorus of living, breathing young people today, clamoring for their right to a stable climate and therefore a hopeful future for themselves and for future generations.

There is no shortage of brilliant, creative, and hardworking young people to listen to. In Paris for the 2015 climate summit, I met a ten-year-old girl boarding the train with her mom. She was carrying a colorful paper gingerbread person as big as she was. "I'm not allowed in the conference," she told me. "So I've made kid-sized dolls to hang around the halls, to remind the negotiators our future depends on them, and we're watching."

We can listen to young people like Lillian, a clear-eyed twelve-year-old from Minnesota who interviewed me for her language arts project on climate change. She told me she had focused on climate because "I made a list of all the things that are wrong with the world. I realized that none of them matter if we don't have a planet. How do I make old people care about climate change?"

Lillian has more power than she may realize, including through sharing her concerns with the adults around her. A study led by Danielle Lawson showed that ten- to fourteen-year-olds who learned about climate change and shared their concerns increased their parents' concern. This was especially the case for daughters. Conservative men were typically the least concerned, and the most resistant group to climate change communication, but they showed the biggest increase in climate concern after listening to their kids. The researchers believe that high levels of parent-child trust, and the children's lack of entrenched political ideology, helped them achieve influence.

Youth is the only universal constituency: Everyone is, or was, a kid. Yet kids can't vote. In 2019, eleven-year-old Lilly Platt requested that her grandfather cast his vote for the EU parliamentary elections according to her wishes that politicians follow the Paris Agreement. He agreed, saying, "Any vote that I make, I don't have to live with the consequences of. You do." Adults can learn a great deal from listening to children with humility.

We can listen to the millions of young people on climate strike in the streets and gathering online around the world, who demand and deserve a world that stops global warming. Their ask is that adults in power now listen to the information we already have from science and act accordingly to safeguard their future.

What Is Humanly Possible?

Climate change is perhaps the most enormous opportunity to test our humanity. Preventing climate breakdown is at the absolute outer limits of what we might just be capable of. It requires us to rewrite our idea of who we are, the language and concepts and traditions we use to braid our identities and cultures together. It requires us to redefine, and then remake, what is humanly possible: what humans are capable of making possible.

I am convinced that now is the most interesting time to be alive. Whether by divine plan or blind chance, it is each and every one of us alive today who happen to be the cast for a decisive moment in human history. Like all generations before us, the impacts of our decisions stretch across space and time in ways we cannot fathom, but we have an unprecedented level of power. Our decisions affect not only ourselves and our families, not only human civilization around the world, today and for millennia to come, but the very preconditions for human civilization: whether, when, and how much rain falls; if crops ripen, or wither and die.

Look around you: your family and neighbors, your coworkers and classmates, leaders at all levels. This is the team. We are it. It's the next decade that's critical in setting the global thermostat, so the fate of the world is in the hands of those of us alive right now. We have this awesome power and responsibility because we are the last stewards of humanity's nearly exhausted carbon budget. We don't have time to wait for someone else. We need to pull together, starting from where we are, cultivating the best of our humanity in this time of crisis, and using it to build and strengthen our individual and collective capacity to do the essential work of regeneration.

We cannot make it the next generation's burden to bear our failures. The true greatest generation has not gone before. We are and must be the greatest generation. Those of us who happen to be alive here and now must make what is necessary possible.

We are in uncharted territory. History cannot be our guide, because there are no examples in history of implementing sweeping global reforms at the scope and scale needed to stabilize the climate. We have to write history to make the necessary changes happen; this is the story we are called to write.

Cathedral Thinking

Each morning I walk to work, I pass a cathedral built almost nine hundred years ago. Thousands of people contributed to building something beautiful they would never live to see finished. History has forgotten their names. Local legend says the cathedral was built by a giant named Finn. In today's twenty-four-hour news cycles and clickbait wars, it seems impossible to imagine a project that would capture and sustain and renew the energy of so many generations.

But in Barcelona today, a great cathedral is rising. A man with a love of geometry and myth imagined something no one else thought possible. Antoni Gaudí helped build the foundation in his lifetime; he hung bags of sand from strings to work out the weight bearing of the arches he would not live to see built. Five generations ago, he imagined the rainbow stained-glass windows whose beauty makes my heart ache. When I visited La Sagrada Familia basilica, the sacred space was full of the whine and buzz of power tools, building the dead man's dream, now also the dream of many who came after him. It is due for completion in the next decade, one hundred years after its architect died. And for untold generations after that, people can stare in wonder at this space that merges nature and mathematics to make something both human and sublime.

Climate stabilization is our generation's cathedral to build. Instead of bricks and mortar, our cathedral must be built from life itself: harnessing regeneration to leave a thriving living world and a good life for ourselves and for future generations as the mark of our time on Earth. We need cathedral thinking to stop climate change: collectively working toward a transcendent purpose we may or may not live to see accomplished but that will outlive us all.

Conclusion

Under the Sky We Make

Colty's baby daughter is now a bright-eyed toddler. She rocks a high ponytail and eighties leg warmers like nobody's business. He recently sent a video of her dancing under a blue summer sky, on the shore of a lake lined with pine trees, delighting in the splish-splash of her pink Crocs against the cool water. In the first few years of her life, about a quarter of the remaining carbon budget has gone up in smoke. She can't hear it yet, but the relentless ticking of the carbon clock is the soundtrack to her whole life.

Meanwhile, Colty and his family have made some changes. They've found a mostly plant-based diet that works for them. They've swapped planned holidays by plane for more local adventures. They've started supporting climate organizations, participating in local demonstrations, and having conversations with friends, family, and coworkers about how they can take climate action. Colty says they have a way better sense of what their political leaders stand for and what their vote means for the climate. Their kids are growing up swimming in the rivers and hiking the hills that surround them. This huge parental schlepp now will be a priceless gift for those kids, who will always have that connection to the places they come from. Colty said he's been amazed by how far he's come, how he views the world differently. He knows where he wants to go, and most of the time, he feels like he's heading in the right direction.

Colty wrote me that "we've moved away from the fear of being Debbie Downer parents who stress out our littles, to the idea that you can't start talking about this critical stuff too early. . . . We see ideas, behaviors, preferences developing SO early in our kiddos and they really are ready to understand the world and to build their own framework for how they will navigate all of this hard stuff."

Colty's insight resonates with my own experiences: The more openly I face the climate crisis from every angle, taking the facts to heart and letting them empower urgency in my actions, the more I feel I'm on the right path. One tiny thing that gives me hope for what humanity can make possible in the coming decade is looking back on how much I personally have changed in the last one, in ways I never would have foreseen, and how much happier I am as a result. Change is hard, but life on the other side of change can be better.

I'm not alone in underestimating how much can change in a decade. Psychologists have found that people readily acknowledge how much they personally have changed in the last ten years, yet do not expect to change further in the future. People tend to regard "the present as a watershed moment at which they have finally become the person they will be for the rest of their lives." Researchers call this "the end of history illusion." But a lot can happen in a decade, for us as individuals and for humanity and this planet.

In this book, we started out facing some brutal facts. Dangerous climate change is not a distant threat. It's part of the world we live in here and now, with impacts that are already heartbreaking

and are poised to escalate to be truly catastrophic. Beyond destabilizing the climate, humans are unraveling the web of life on
Earth; we are the asteroid currently set on a collision course.

But humanity is a conscious asteroid; we can choose to change
course. Importantly, it's the Exploitation Mindset, not some inevitable fact of human nature, that is driving us to the brink of devastation. Science has documented the problems in excruciating
detail and has a plethora of solutions to stabilize the climate and
the living world waiting in the wings, but science itself won't save
us. What will is a wave of people rising up and joining together to
weave care, increased resilience, and decreased harm throughout
our lives and societies. Making these changes in how we treat
nature and one another is how we will make the transition to a
fossil-free and regenerative world. The fate of the planet and its
inhabitants rests with our stewardship today.

Our responsibility in this particular moment of life on Earth is
a whole lot to take in. But here we are. In Part 2, we looked at how
to face All the Climate Feels that accompany living through this
challenging but also transformative time, and how to help one another get through it together through harnessing the power of our
feelings and community. We'll each need to draw on our personal
core values and our shared humanity to identify and pursue a
North Star of purpose in this warming world. When (not if) the
going gets tough, our sense of meaning from choosing and doing
this work, and the support and joy from our relationships, will
help us keep putting one foot in front of the other.

Bolstered by facts and feelings, in Part 3 we got down to work.
We saw that our responsibility and urgency scale with our privilege and power. We also examined how the Exploitation Mindset
is currently woven through our lives, tangled up in our economic,
political, and cultural systems, and built into sectors from transportation to energy to food. Now that we've seen these things, we

can't unsee them. It is going to be a truly heroic job to start undoing all the harm from the well-worn grooves of exploitation and point toward regeneration throughout these systems instead. No one can do it all, but everyone can do something with their unique gifts. I hope you've found ways to start putting yours to work.

As for me, I started out telling stories of my family's past, to show how everyone's history, and ultimately the story of humanity, is tied up with the earth and the sky we're making now. Thinking about myself as part of a family, I realize I am my ancestors' future generation, already far, far down the branches of my family tree. Many of my opportunities and challenges are predicated on the choices they made.

I've cut my flying more than 90 percent, but I haven't yet given up all my "love miles" to see my family and friends across the ocean. I have two nephews in college whom I love to take canoeing, and a fairy goddaughter who loves Taylor Swift and is scared of bugs. Some of the carbon I release when I fly back to California to see them will remain in the atmosphere long after their children's children's children are gone.

Those of us alive today have the opportunity and the obligation to be a good ancestor, not just to our own descendants but to the continuation of humankind, to the renewal and regeneration of the values that celebrate and elevate our humanity. We are called to create, live in, and pass on a world where each person is treated with dignity and has the material and social fundamentals they need for a good life. A world where people can continue to grow sufficient food. A world where life is thriving in the oceans and on land. A world where life is no longer threatened by climate change. How far down the branches of the family tree we are, how many more branches go forward, depends so much on what we do today.

We are living under the sky we make. We have made it with

our behavior, and we can remake it that way too. Rather than extrapolating the current darkness to despair, we need to work backward from the world we want to find a path there, in order to create a world we love rather than one we fear.

Will humanity manage to throw all we've got into actually fixing the world we're breaking? Will we accept our responsibility to get at least half the job done by 2030, or will we miss our last clear chance to avoid catastrophic climate change?

I expect to live long enough to find out. Statistically, I expect to die on May 17, 2062, at the age of 84.4. Based on our collective human choices in the next decade, I will die either on a planet stabilizing around 1.5°C warmer than the one of my ancestors, or on a planet that's already blown past the 1.5°C and 2°C limits on the way to the hellscape of 3°C or beyond.

If humans make the earth basically uninhabitable for many of the plants and animals and civilizations alive today, no matter our other accomplishments, I believe we will have failed as a species.

When I look back on my life, I expect to assess my personal meaning in terms of the love I have shared with those closest to my heart. I will evaluate my contribution to the world in terms of how much carbon I helped keep out of the atmosphere, because that influence will far, far outlast me and anything else I might leave behind. When I die, I will judge a great deal of the success and meaning of the whole human endeavor by what kind of ancestors we were, measured by the legacy we leave of parts per million of carbon dioxide in the atmosphere. This is ultimately the language in which the story of the fate of humanity and planet Earth is written: these molecules we send skyward to be taken up into leaf, and shell, and stone.

TLDR (Too Long, Didn't Read)

You're busy. I get it. Here are the most important takeaways and key actions from each chapter.

Introduction: Science Won't Save Us

- It's warming, it's us, we're sure, it's bad, we can fix it.

- We basically have the tech we need to stabilize the climate.

- We need to clarify our values and shift our mindsets and actions in line with what the science tells us is necessary to stop climate and ecological breakdown and preserve humanity.

- To face the climate crisis, we need to harness facts, feelings, and action.

Chapter 1: Carbon Is Forever

- To meet the 1.5°C climate goal, we need to cut global carbon pollution at least in half by 2030. High emitters should cut more than half.

- Because some carbon lasts essentially forever in our atmosphere, carbon pollution has to go *all the way to zero* to stabilize the climate. This means we have to quickly replace fossil fuels with clean energy, and switch from polluting to regenerative agriculture.

Chapter 2: We're the Asteroid

- Human exploitation of the land, water, and oceans is killing a lot of life on Earth. This is both morally wrong and stupid; we need nature to survive and to feel our most human.

Chapter 3: Uprooting Exploitation, Sowing Regeneration

- The root cause of both the climate and ecological emergencies is what I call the Exploitation Mindset: thinking that some humans should dominate other humans and that humans in general should dominate nature.

- We need to learn to live well within the material limits of the biosphere. To do so, we need to redefine what a good life looks like.

- Three principles to guide us in uprooting the Exploitation Mindset and cultivating the Regeneration Mindset are: respect and care for people and nature; reduce harm at its source; and build resilience, the ability to recover from setbacks.

Chapter 4: Sink into Your Grief

- Climate and ecological breakdown have caused, are causing, and will cause irreplaceable losses of things we love, and that's really sad.

- We have to acknowledge and make space for grief to honor what we're losing, and to motivate us to fight for what's left.

Chapter 5: Making Meaning in a Warming World

- Meaning is created from actions and relationships that align with our core values.

- Working to undo climate change provides a crucible to create meaning in our lives: it provides a role to play in an epic story, a purpose in pursuing goals chosen to serve our core values, and a way of mattering to others and the world.

- Maximize meaning, minimize carbon, is one measure of a life well lived.

- Look for work that combines what you love, are good at, and think is fun. The climate needs everyone's talents.

- Having a child has a big, long-term climate impact. But stabilizing the climate hinges on decarbonizing in the next decade. As long as you're on board for that, use your core values to decide whether having a child is one of your central goals.

Chapter 6: Face Your Fears

- It can be hard and overwhelming to work on a problem as big as climate change. That's normal and okay. We don't get to give up or turn into assholes, though.

- Recognize the five stages of radical climate acceptance (Ignorance, Avoidance, Doom, All the Feels, and Purpose). Use this awareness to help you navigate them more skillfully.

- Building and cultivating a supportive community is a key strategy to boost your capacities and increase your resilience.

Chapter 7: Get Angry

- The fossil fuel industry organized a disinformation campaign to falsely cast doubt on the climate harm their products cause, and to delay necessary shifts from dirty to clean energy. I do not know how they sleep at night. Use anger at injustice to power action.

- Facts are not enough to change someone's mind if their identity rests in a different belief.

- But! Most Americans believe facts and trust scientists. (Thanks, peeps, love you too!)

Chapter 8: Climate Change Isn't Fair

- Climate pollution is distributed extremely unfairly. Overall, the United States and Europe, high-income individuals, and people alive in the last four decades have polluted way more than their fair share.

- Both individual and collective actions, in both private and professional life, are needed to reduce emissions toward zero fast.

- Everyone's personal climate budget needs to be 2.5 tons per year by 2030. This will require both system change to meet human needs without producing climate pollution, and eliminating overconsumption (lifestyle changes) for high emitters.

- The personal actions that cut climate pollution fast are to go flight-, car-, and meat-free. Start with the one that feels most feasible for you; if you can't totally go without, aim to cut your consumption today at least in half.

- Climate privilege means the more power and privilege I have, the more responsibility I need to take. Others having even more responsibility does not absolve me of my own.

- Change happens by internalizing the urgency of the climate crisis and accepting personal responsibility, along with seeing others around us take action. Walking the talk is important.

Chapter 9: Slowing Down and Staying Grounded

- High emitters must reduce our carbon overconsumption to stabilize the climate.

- If you fly, it's your most intense source of climate pollution. Stay on the ground.

- Cars are most Americans' largest source of climate pollution. Go car-free.

- Support regulations and standards to electrify everything on zero-carbon energy.

- Consider getting solar panels and making your home more energy efficient. Reduce your use of everything that heats and cools (air conditioning, water heaters, etc.) as much as possible.

Chapter 10: Food Shouldn't Come from a Factory

- Eat a plant-based diet. Try going vegan for a month. If you eat animal products, eat ones raised well, and no more than: one glass of milk or one ounce of cheese per day; two eggs and two servings of chicken and fish per week; two burgers per month. Eat more (nontropical) fruits, vegetables, legumes, and nuts. Limit coffee and chocolate (sorry!).

- Support farm subsidy reform to stop paying polluters and widening income inequality.

- Grow some food wherever you can.

- Keep cats indoors; they kill lots of birds.

- Protect remaining nature. Restore nature where it's degraded. Support rewilding in collaboration with local communities. Support circular resource management that turns waste into resources.

Chapter 11: Values and Costs

- Become an economic "growth agnostic": Remember that growth in well-being, not GDP, is the true purpose of the economy.

- Support ending fossil fuel subsidies in all their forms. Make sure they're replaced with fair policies that protect the vulnerable.

- Recognize that the market alone cannot stabilize the climate, though a price on emitting carbon can help somewhat in the short term. Ultimately, the carbon emission target needed is zero, so the correct price for carbon is infinite: a ban prohibiting fossil exploration, construction, supply, production, and financing.

- Stop all forms of investment in fossil energy immediately. Direct that money and more to invest in clean energy, especially decentralized community energy. Prepare to shut down existing fossil infrastructure early.

- Break up with your climate-destroying mega-bank and credit card company. Switch to a credit union.

- Support or start campaigns for your school, workplace, pension, city, etc. to divest from fossil fuels.

- Pressure your employer to develop and implement a plan to decarbonize their operations and industry.

- Offsets don't work. Donate 10 percent of your income, or as much as you can, to groups working for change without considering it to absolve you of your climate pollution.

Chapter 12: The Personal Is Political

- Vote for women and for candidates with good climate scores who have taken the No Fossil Fuel Money Pledge. Write and call your rep directly with your climate concerns.

- Get active in climate organizations and social movements, create media attention, and participate in nonviolent demonstrations and creative actions.

- Demand corporate and government leaders deliver a plan for rapidly reducing climate pollution within their domain toward zero. Make them focus on reducing and preventing, not removing, their climate pollution.

- Cultivate fair processes to involve citizens in climate deliberation and decision-making.

- Support just transitions for workers from industries that need to decline to meet climate goals.

- Support a free and independent press; subscribe to (and pay for) good climate journalism. Pressure media to stop accepting ads from climate destroyers.

- Support legal cases for climate; they're being heard and starting to win.

Chapter 13: Being a Good Ancestor

- It's the next decade that's critical in setting the global thermostat, so the fate of the world is in the hands of those of us alive right now.

- Find bright spots where climate action is working and help spread them.

- Expect and learn from failures along the way.

- To be a good ancestor, we can forget hypothetical future generations. Listen with humility to today's young people demanding their right to a stable climate.

Conclusion: Under the Sky We Make

- A lot can happen in a decade, and the one ahead is critical.

- We are living under the sky we make; let's make it good.

Acknowledgments

I am overwhelmed with gratitude for the incredible support I've received from colleagues, friends, and family to help make this book a reality.

Thank you to my fierce agent Anna Sproul-Latimer. You are very good at what you do, and you make me a better writer, author, and person. Plus you make the best statement resin jewelry. I'm so glad to be part of the Neon Literary dream team, and grateful to Kent Wolf for comments and support.

Thank you to my editor, Michelle Howry, for your unflagging encouragement, reassurance, and guidance. Thanks to Ashley Di Dio, Brennin Cummings, Kristen Bianco, Sydney Cohen, and to the whole team at Putnam for helping bring this book to life.

I'm grateful to Emily Boyd and my colleagues at LUCSUS for supporting my leave of absence to finish this book.

Thank you to all the hardworking researchers who have tirelessly quantified our predicament, and the journalists who report it. I am deeply indebted to your work on which I draw here.

I am truly grateful to the generous colleagues who contributed their time and expertise to critique the text, share references (the record was nine PDFs sent to support one sentence, which ultimately got deleted anyway), discuss and argue over scientific nuances, and help me refine my arguments as this book was under development.

Thank you to chapter experts Kevin Anderson, Shahzeen Attari, Elena Bennett, Emily Cassidy, Aleh Cherp, Søren Faurby, Taryn Fransen, Ben Franta, Aarne Granlund, Eve Hinckley, Frida Hylander, Diana Ivanova, Kirsti Jylhä, Peter Kalmus, Stephanie La Hoz Theuer, Giulio Mattioli, Geneviève Metson, Rebecca Miller, Julia Mosquera, Kristian Steensen Nielsen, Allison Perrigo (what a gift to have a reviewer whose scientific nitpickings can make me LOL), Ludwig Bengtsson Sonesson, Johannes Stripple, and Parke Wilde.

Thank you to the topic experts, who provided studies, data, explanations, clarifications, and fact-checks: John Abatzoglou, Bill Anderegg, Almut Arneth, Jill Bible, Chad Boda, Kate Brauman, Mike Cahill, Ken Caldeira, Stuart Capstick, Kim Cobb, Edward Collins, Chris Doughty, Nils Droste, Michael Eisenstein, Karlheinz Erb, Eléonore Fauré, Jon Foley, Lisa Friedman, Sara Gabrielsson, Stefan Gössling, Emma Johansson, Jacob Johansson, Tobias Kuemmerle, Eric Lyman, Sophie Marjanac, Bill McKibben, Ina Möller, Frances Moore, Alexander James Mustill, Tobias Nielsen, Camille Parmesan, Glen Peters, Ken Rice, Julia Steinberger, Sarah Ullström, Mehana Vaughan, Richard Waite, and Christine Wamsler.

Thank you to Darin Wahl for the whiteboard brainstorming that led to naming the Exploitation and Regeneration Mindsets, and to Denise Young for brainstorming subtitles when I was beyond stuck. Thank you, Atticus Pinzon-Rodriguez, for your Photoshop and meme wizardry.

Thank you especially to colleagues and friends who read all or most of the whole dang thing. Charlie Wilson made me fundamentally rethink more than I wanted to, as usual; Seth Wynes gave fresh insight and wordsmithing, as usual. Vance Wagner gave inspiration at just the right time. Thank you to Ritik Dholakia for helpful conversation and feedback.

I'm (mostly) grateful to #ClimateTwitter, especially people like @cityatlas and @CostaSamaras, who were unfailingly helpful and often made me laugh.

Thank you to the people who helped me get things right: Emma Copley Eisenberg, for your *Esquire* article that convinced me to get fact-checkers, and Miguel Salazar, research director at *The Nation,* for your kindness in setting me on the fact-checking path. Thank you to stellar fact-checkers Noah Flora and Jocelyn Timperley. Thank you to Klara Winkler for research assistance and to assistance from My Knutsson in reference formatting.

All remaining errors are my own.

I am honored to have friends who took the time to read and respond to drafts, including Linda Voris. Thank you to my beloved book club—Martin Nilsson, Fernanda Åkerman, Rui Martins, Ezra Chomak, Simon Rose, Kerstin Wolf (repeatedly!), and Zhanna Kryukova—for reading and critiquing the first full draft. It was a thrill to write for you.

Thank you to the friends who have shared the whole journey with me and shaped how and why I wrote this book: Lucy Goddard Kalanithi, Meg Pearson, Colty Tipton-Johnson, Rebecca Lewis, Jennie Clare, Cara Harrison.

Thank you to the Real Live Authors and Editors who were kind enough to offer me advice and encouragement: Caitlin Doughty, Thomas Hynes, Mark Fischetti, Torill Kornfeldt (I still have my inspirational Post-it "Write a dry, boring book"), and Hope Jahren (I will never forget your advice, "The way to write a book is like the way to have an orgasm—however you can")!

Thank you to my family. I appreciate the lifelong support of my parents, including my dad reading a draft with his sharp eye for detail, and my mom with her sharp mind for argument. I'm grateful to my sister and her family for their love. Thank you to my aunt Judy Glock for being the family historian to help keep Clara's story alive. Finally, thank you to Simon for everything. I'm so glad we swiped right.

Notes

Introduction

2 **Only 10 percent don't believe:** M. Goldberg et al., "For the First Time, the Alarmed Are Now the Largest of Global Warming's Six Americas," Yale Program on Climate Change Communication, January 16, 2020, https://climatecommunication.yale.edu/publications/for-the-first-time-the-alarmed-are-now-the-largest-of-global-warmings-six-americas/.

4 **just five key facts:** Thanks to Jon Krosnick and Susan Hassol for first articulating these points. See Kimberly Nicholas, "Climate Science 101: Five Things Everyone Needs to Know," http://www.kimnicholas.com/climate-science-101.html. For a scientific review of the evidence behind these points, see Seth Wynes and Kimberly Nicholas, "Climate Science Curricula in Canadian Secondary Schools Focus on Human Warming, Not Scientific Consensus, Impacts or Solutions," *PLoS One* 14, no. 7 (2019): e0218305 (table 1), https://doi.org/10.1371/journal.pone.0218305.

4 **we have the tech we need:** S. J. Davis et al., "Net-Zero Emissions Energy Systems," *Science* 360, no. 6396 (June 29, 2018): figure 2, https://doi.org/10.1126/science.aas9793; White House, "United States Mid-Century Strategy for Deep Decarbonization," November 2016, 30, https://unfccc.int/files/focus/long-term_strategies/application/pdf/us_mid_century_strategy.pdf.

6 **most of the world's problems:** Thomas Wiedmann et al., "Scientists' Warning on Affluence," *Nature Communications* 11, no. 3107 (2020), https://doi.org/10.1038/s41467-020-16941-y.

8 **We need a new mindset:** My inspiration to focus on personal transformations of the mindsets in which systems are embedded to support broader transformation comes from sources including Donella Meadows, "Leverage Points: Places to Intervene in a System," Sustainability Institute, 1999, 2–3, http://donellameadows.org/wp-content/userfiles/Leverage_Points.pdf; Intergovernmental Science-Policy Platform on Biodiversity and Ecosystem

Services (IPBES), *Summary for Policymakers, Global Assessment Report on Biodiversity and Ecosystem Services* (Bonn, Germany: IPBES Secretariat, 2019), 14; Karen O'Brien, "Is the 1.5°C Target Possible? Exploring the Three Spheres of Transformation," *Current Opinion in Environmental Sustainability* 31 (2018): 153–60, https://doi.org/https://doi.org/10.1016/j.cosust.2018 .04.010.

Chapter 1

14   **next *three hundred* generations:** I took the average length of a generation as thirty years (Donn Devine, "How Long Is a Generation?" *Ancestry*, September–October 2005, 51–53): 10,000 years for the long-lived fraction of atmospheric carbon, as explained shortly, divided by 30 years equals 333 generations.

15   **all known life-forms:** S. E. Gould, "Shine on You Crazy Diamond: Why Humans Are Carbon-Based Lifeforms," *Scientific American*, November 11, 2012, https://blogs.scientificamerican.com/lab-rat/shine-on-you-crazy -diamond-why-humans-are-carbon-based-lifeforms/.

15   **climate is what you expect:** Garson O'Toole, "The Climate Is What You Expect; the Weather Is What You Get," Quote Investigator, June 12, 2012, https://quoteinvestigator.com/2012/06/24/climate-vs-weather/.

16   **75 percent of heat-trapping emissions:** Intergovernmental Panel on Climate Change (IPCC), "Summary for Policymakers," in *Climate Change 2014: Mitigation of Climate Change. Contribution of Working Group III to the Fifth Assessment Report of the Intergovernmental Panel on Climate Change*, ed. O. Edenhofer et al. (Cambridge, UK: Cambridge University Press, 2014), figure SPM.1.

16   **continents and oceans sing:** IPCC, "Summary for Policymakers," in *Climate Change 2013: The Physical Science Basis. Contribution of Working Group I to the Fifth Assessment Report of the Intergovernmental Panel on Climate Change*, ed. T. F. Stocker et al. (Cambridge, UK: Cambridge University Press, 2013), figure SPM.1, 6.

16   **Thoreau's Concord journal:** Richard B. Primack and Abraham J. Miller-Rushing, "Uncovering, Collecting, and Analyzing Records to Investigate the Ecological Impacts of Climate Change: A Template from Thoreau's Concord," *BioScience* 62, no. 2 (2012): 170–81, https://doi.org/10.1525/bio .2012.62.2.10.

16   **approximately 1°C above:** IPCC, "Summary for Policymakers," in *Global Warming of 1.5°C. An IPCC Special Report on the Impacts of Global Warming of 1.5°C Above Pre-Industrial Levels and Related Global Greenhouse Gas Emission Pathways, in the Context of Strengthening the Global Response to the Threat of Climate Change, Sustainable Development, and Efforts to Eradicate Poverty*, ed. V. Masson-Delmotte et al. (Geneva, Switzerland: IPCC, 2018), statement A.1.

17   **CO_2 has increased 40 percent . . . increase their acidity 26 percent:** IPCC, *Climate Change 2014: Synthesis Report* (Geneva, Switzerland: IPCC, 2015), 4, 44.

17   **pinpoint the source:** O. J. Schneising et al., "Anthropogenic Carbon Dioxide Source Areas Observed from Space: Assessment of Regional Enhancements and Trends," *Atmospheric Chemistry and Physics* 13, no. 5 (2013): 2445–54, https://doi.org/10.5194/acp-13-2445-2013.

17   **consistent with what physics predicts:** K. E. Trenberth et al., "Frequently Asked Question 3.1: How Are Temperatures on Earth Changing?" in chapter 3 of IPCC, *Climate Change 2007: The Physical Science Basis. Contribution of Working Group I to the Fourth Assessment Report of the Intergovernmental Panel on Climate Change*, ed. S. Solomon et al. (Cambridge, UK: Cambridge University Press, 2007), https://archive.ipcc.ch/publications _and_data/ar4/wg1/en/faq-3-1.html.

17   *more than 100 percent* **of warming:** Zeke Hausfather, "Analysis: Why Scientists Think 100% of Global Warming Is Due to Humans," Carbon Brief, December 13, 2017, https://www.carbonbrief.org/analysis-why-scientists -think-100-of-global-warming-is-due-to-humans.

17   **About 30 percent of the cumulative carbon:** Global Carbon Project, "Global Carbon Budget 2019" (PowerPoint presentation, 2019), slide 54, https://www.globalcarbonproject.org/carbonbudget/19/presentation.htm.

17   **clearing vegetation and disturbing soil:** T. F. Stocker et al., "Technical Summary," in *Climate Change 2013*, 96, https://www.ipcc.ch/report/ar5 /wg1/technical-summary/.

17   **a quarter of our total climate pollution:** IPCC, "Summary for Policymakers," in *Climate Change and Land: An IPCC Special Report on Climate Change, Desertification, Land Degradation, Sustainable Land Management, Food Security, and Greenhouse Gas Fluxes in Terrestrial Ecosystems*, ed. P. R. Shukla, et al. (2019), figure SPM.1, https://www.ipcc.ch/srccl/chapter /summary-for-policymakers/.

18   **86 percent of the carbon:** "Global Carbon Budget 2019," slide 45.

18   **36.8 billion tons:** Pierre Friedlingstein et al., "Global Carbon Budget 2019," *Earth System Science Data* 11, no. 4 (2019): 1811, https://doi.org /10.5194/essd-11-1783-2019.

18   **cleans up a bit over half:** Stocker et al.,"Technical Summary," 96.

18   **Slightly more than a quarter:** "Global Carbon Budget 2019," slide 45.

18   **someone keeps stealing your bricks:** Elizabeth Kolbert, "The Darkening Sea," *The New Yorker*, November 20, 2006, 69.

18   *ten thousand years later:* Stocker et al., "Technical Summary," 96; Mason Inman, "Carbon Is Forever," *Nature Climate Change* 1 (2008): 156–58, https://www.nature.com/articles/climate.2008.122.pdf.

19   **more carbon in three days:** Terry Gerlach, "Volcanic Versus Anthropogenic Carbon Dioxide," *EOS, Transactions, American Geophysical Union* 92, no. 24 (2011): 201–8, https://doi.org/10.1029/2011EO240001.

19   **hundreds of thousands of times faster:** Holli Riebeek, "The Carbon Cycle," NASA Earth Observatory, June 16, 2011, https://earthobservatory .nasa.gov/features/CarbonCycle.

19   **280 parts per million (ppm):** "Global Carbon Budget 2019," slide 54.

19   **passed 300 ppm:** D. Luthi et al., "High-Resolution Carbon Dioxide Concentration Record 650,000–800,000 Years Before Present," *Nature* 453, no. 7193 (2008): figure 2, https://doi.org/10.1038/nature06949.

19   **In 2019, it averaged 411 ppm:** Pieter Tans, NOAA/GML (www.esrl .noaa.gov/gmd/ccgg/trends/) and Ralph Keeling, Scripps Institution of Oceanography (scrippsco$_2$.ucsd.edu/). Mauna Loa CO_2 annual mean data for 2019, downloaded August 1, 2020.

19   **started rising rapidly . . . about 2.3 ppm per year:** Rebecca Lindsey, "Climate Change: Atmospheric Carbon Dioxide," NOAA Climate, February

20, 2020, https://www.climate.gov/news-features/understanding-climate/climate-change-atmospheric-carbon-dioxide.

19 **The "safe" level is 350 ppm:** J. Hansen et al., "Target Atmospheric CO_2: Where Should Humanity Aim?" *Open Atmospheric Science Journal* 2 (2018): 217–31; Johan Rockström et al., "A Safe Operating Space for Humanity," *Nature* 461, no. 7263 (2009): 472–75, http://dx.doi.org/10.1038/461472a.

19 **blew past in 1987:** Lindsey, "Climate Change."

19 **sluggish politics have always lagged:** David Biello, "Climate Negotiations Fail to Keep Pace with Science," *Scientific American,* December 7, 2011, https://www.scientificamerican.com/article/climate-negotiations-fail/.

20 **"prevent dangerous anthropogenic [human-caused] interference":** United Nations, *United Nations Framework Convention on Climate Change* (1992), article 2, https://unfccc.int/resource/docs/convkp/conveng.pdf.

20 **A long political process:** Reto Knutti et al., "A Scientific Critique of the Two-Degree Climate Change Target," *Nature Geoscience* 9, no. 1 (2015): 13–18, https://doi.org/10.1038/ngeo2595.

20 **"well below 2°C":** United Nations, *Paris Agreement* (Paris, France, 2015), article 2, https://unfccc.int/files/essential_background/convention/application/pdf/english_paris_agreement.pdf.

20 **"allow ecosystems to adapt naturally":** United Nations, *United Nations Framework Convention on Climate Change,* article 2.

20 **more than *one hundred times* faster:** Scripps Institution of Oceanography, "Carbon Dioxide at Mauna Loa Observatory Reaches New Milestone: Tops 400 ppm," May 10, 2013, https://scripps.ucsd.edu/news/7992.

21 **squeezed off the mountaintops:** Eben H. Paxton et al., "Collapsing Avian Community on a Hawaiian Island," *Science Advances* 2, no. 9 (2016), https://10.1126/sciadv.1600029.

21 **"assisted colonization":** T. van Dooren, "Moving Birds in Hawai'i: Assisted Colonisation in a Colonised Land," *Cultural Studies Review* 25, no. 1 (2019): 41–64, https://doi.org/10.5130/csr.v25i1.6392.

21 **half of all species studied:** Camille Parmesan, quoted in "Climate Change 'Forcing Species to Move,'" Phys.org, February 10, 2016, https://phys.org/news/2016-02-climate-species.html.

21 **study led by Deepak Ray:** D. K. Ray et al., "Climate Change Has Likely Already Affected Global Food Production," *PLoS One* 14, no. 5 (2019): e0217148, https://doi.org/10.1371/journal.pone.0217148.

21 **yields for four crops:** David B. Lobell, Wolfram Schlenker, and Justin Costa-Roberts, "Climate Trends and Global Crop Production Since 1980," *Science* 333, no. 6042 (2011): 616–20, https://doi.org/10.1126/science.1204531.

21 **crop productivity in almost a quarter:** Intergovernmental Science-Policy Platform on Biodiversity and Ecosystem Services (IPBES), *Summary for Policymakers, Global Assessment Report on Biodiversity and Ecosystem Services* (Bonn, Germany, 2019), statement A3, https://doi.org/10.5281/zenodo.3553579.

22 **increased the income gap:** Noah S. Diffenbaugh and Marshall Burke, "Global Warming Has Increased Global Economic Inequality," *Proceedings of the National Academy of Sciences of the United States of America* 116, no. 20 (2019): 9808–13, https://doi.org/10.1073/pnas.1816020116.

22 **review of 3,280 research papers:** C. Mora et al., "Broad Threat to Humanity from Cumulative Climate Hazards Intensified by Greenhouse Gas Emissions," *Nature Climate Change* 8, no. 12 (2018): 1062.

22 **"blatantly inadequate":** United Nations Environment Programme (UNEP), *Emissions Gap Report 2019* (Nairobi: UNEP, 2019).

22 **our atmosphere is like a bathtub:** Climate Interactive, "Climate Bathtub Simulation," https://www.climateinteractive.org/tools/climate-bathtub -simulation/.

23 **something like 90 percent . . . less than 70 percent:** Carbon budget calculations for 1.5°C are derived from Global Carbon Project data, https:// www.globalcarbonproject.org/carbonbudget/19/visualisations.htm.

24 **at today's rate, before 2030:** IPCC, *Global Warming of 1.5°C*, statement C1.3.

24 **all the way to zero:** IPCC, *Global Warming of 1.5°C*, statement C1.2 and figure SPM.3A.

25 **latest IPCC assessment warns:** Stocker et al., "Technical Summary," 98, box TS.7.

25 **Life expectancy more than doubled:** Max Roser, Esteban Ortiz-Ospina, and Hannah Ritchie, "Life Expectancy," Our World in Data, 2013, https://ourworldindata.org/life-expectancy.

26 **under the age of twenty-five:** Hannah Ritchie, "Age Structure," Our World in Data, 2019, https://ourworldindata.org/age-structure.

26 **warming that could happen this century:** Zeke Hausfather, "Explainer: How 'Shared Socioeconomic Pathways' Explore Future Climate Change," Carbon Brief, April 19, 2018, https://www.carbonbrief.org/ex plainer-how-shared-socioeconomic-pathways-explore-future-climate-change. I also used the IIASA Scenario Explorer thanks to a suggestion from Joeri Rogelj on Twitter: Daniel Huppmann et al., "IAMC 1.5°C Scenario Explorer and Data Hosted by IIASA," Integrated Assessment Modeling Consortium and International Institute for Applied Systems Analysis, 2019, https://doi .org/10.5281/zenodo.3363345.

26 **era of committed climate change:** K. R. Smith et al., "Human Health: Impacts, Adaptation, and Co-benefits," in *Climate Change 2014: Impacts, Adaptation, and Vulnerability. Part A: Global and Sectoral Aspects. Contribution of Working Group II to the Fifth Assessment Report of the Intergovernmental Panel on Climate Change*, ed. C. B. Field et al. (Cambridge, UK: Cambridge University Press, 2014), figure 11-6, 735, https://www.ipcc.ch /report/ar5/wg2/human-health-impacts-adaptation-and-co-benefits/.

26 **the IPCC warns that:** Smith et al., "Human Health," figure 11-6, 735.

27 **an Ice Age is about 4°C colder:** E. Jansen et al., "FAQ 6.2: Is the Current Climate Change Unusual Compared to Earlier Changes in Earth's History?" in chapter 6, "Palaeoclimate," of *Climate Change 2007*, https://archive .ipcc.ch/publications_and_data/ar4/wg1/en/faq-6-2.html.

27 **During the last Ice Age:** "Facts About Glaciers," National Snow and Ice Data Center, n.d., https://nsidc.org/cryosphere/glaciers/quickfacts.html.

27 **Katharine Hayhoe points out:** Devin Thorpe, "Everything You Ever Wanted to Know About Climate Change but Were Afraid to Ask," *Forbes*, December 9, 2019, https://www.forbes.com/sites/devinthorpe/2019/12/09 /everything-you-ever-wanted-to-know-about-climate-change-but-were-afraid-to -ask/#25f004da3a60.

27 **2.5 billion smallholder farmers:** "Smallholders, Food Security, and the Environment," International Fund for Agricultural Development and United Nations Environment Programme (2013), 8, https://www.ifad.org/documents /38714170/39135645/smallholders_report.pdf/.

28 **at current emissions rates:** IPCC, *Global Warming of 1.5°C*, statement A1.

28 **three times as much habitat . . . cleaner air:** IPCC, *Global Warming of 1.5°C*, statement B3.1, statement C.1.2.

28 **repeated and increasing bleaching worldwide:** T. P. Hughes et al., "Global Warming and Recurrent Mass Bleaching of Corals," *Nature* 543, no. 7645 (2017): 373, https://doi.org/10.1038/nature21707.

28 **"wake-up call that we are simply out of time":** Katharine Bagley, "From Mass Coral Bleaching, a Scientist Looks for Lessons," *Yale Environment 360*, April 27, 2016, https://e360.yale.edu/features/from_mass_coral _bleaching_scientist_looks_for_lessons_kim_cobb_el_nino.

29 **"But we're going to be next":** Kim Cobb, "The Flying-Less Movement Webinar," starting at 23:50, quote at 24:54–25:42, October 30, 2019, https:// www.youtube.com/watch?v=yWas9U4Q_BM&t=1543s.

29 **only 10 to 30 percent:** IPCC, *Global Warming of 1.5°C*, statement B4.2.

29 **virtual elimination of coral reefs:** IPCC, *Global Warming of 1.5°C*, statement B4.2.

29 **2060 under current climate policies:** Ed Hawkins, "When Will We Reach 2°C?" Climate Lab Book: Open Climate Science, April 15, 2014, https://www.climate-lab-book.ac.uk/2014/when-will-we-reach-2c/.

29 **3.6 billion people:** IPCC, *Global Warming of 1.5°C*, Table 3.4.

29 **US residents would feel:** M. C. Fitzpatrick and R. R. Dunn, "Contemporary Climatic Analogs for 540 North American Urban Areas in the Late 21st Century," *Nature Communications* 10, no. 1 (2019): figure 4a, https://fitzlab .shinyapps.io/cityapp/ and https://doi.org/10.1038/s41467-019-08540-3.

29 **more than half of today's wine regions:** I. Morales-Castilla et al., "Diversity Buffers Winegrowing Regions from Climate Change Losses," *Proceedings of the National Academy of Sciences of the United States of America* 117, no. 6 (2020): 2864–69, https://doi.org/10.1073/pnas.1906731117.

29 **about 3.5°C by 2100:** UNEP, *Emissions Gap Report 2019*.

29 **average human would experience more than twice:** C. Xu et al., "Future of the Human Climate Niche," *Proceedings of the National Academy of Sciences of the United States of America* 117, no. 21 (2020): 11350–55, https://doi.org/10.1073/pnas.1910114117.

30 **at least 3 million years ago:** K. D. Burke et al., "Pliocene and Eocene Provide Best Analogs for Near-Future Climates," *Proceedings of the National Academy of Sciences of the United States of America* 115, no. 52 (2018): 13288–93, https://doi.org/10.1073/pnas.1809600115.

30 **camels roamed north of the Arctic:** N. Rybczynski et al., "Mid-Pliocene Warm-Period Deposits in the High Arctic Yield Insight into Camel Evolution," *Nature Communications* 4 (2013): 1550, https://doi.org/10.1038 /ncomms2516.

30 **continues to fail:** UNEP, *Emissions Gap Report 2019*, 27.

30 **more warming is not out of the question:** Robinson Meyer, "Are We Living Through Climate Change's Worst-Case Scenario?" *The Atlantic*, Janu-

ary 15, 2019, https://www.theatlantic.com/science/archive/2019/01/rcp-85
-the-climate-change-disaster-scenario/579700/.

30 **85 percent of today's beloved wine-growing regions:** I. Morales-
Castilla et al., "Diversity Buffers Winegrowing Regions," 2864–69.

30 **much of life at sea and on land will be dead or dying:** IPCC,
Climate Change 2014: Synthesis Report; M. C. Urban, "Accelerating Extinction Risk from Climate Change," *Science* 348, no. 6234 (2015): 571–73,
https://doi.org/10.1126/science.aaa4984.

30 **"sustained food supply disruptions globally":** IPCC, "Summary for
Policymakers," *Climate Change and Land,* figure SPM.2.

30 **avocados will be toast:** D. B. Lobell, K. N. Cahill, and C. B. Field, "Historical Effects of Temperature and Precipitation on California Crop Yields,"
Climatic Change 81, no. 2 (2007): 187–203, https://doi.org/10.1007/s10584
-006-9141-3; D. B. Lobell et al., "Impacts of Future Climate Change on California Perennial Crop Yields: Model Projections with Climate and Crop Uncertainties," *Agricultural and Forest Meteorology* 141, nos. 2–4 (2006): figure 3,
https://doi.org/10.1016/j.agrformet.2006.10.006.

30 **basic staple grains:** T. A. Pugh et al., "Climate Analogues Suggest Limited
Potential for Intensification of Production on Current Croplands Under Climate
Change," *Nature Communications* 7 (2016): 12608, https://doi.org/10.1038
/ncomms12608; N. W. Arnell et al., "Global and Regional Impacts of Climate
Change at Different Levels of Global Temperature Increase," *Climatic Change*
155, no. 3 (2019): 377–91, https://doi.org/10.1007/s10584-019-02464-z.

30 **activities of daily living impossible:** Smith et al., "Human Health,"
section 11.8.1.

31 **likens tipping points to a game of Jenga:** Robert McSweeney, "Explainer: Nine 'Tipping Points' That Could Be Triggered by Climate Change,"
Carbon Brief, February 10, 2020, https://www.carbonbrief.org/explainer
-nine-tipping-points-that-could-be-triggered-by-climate-change.

31 **twice as much carbon in soils:** P. Ciais et al., "Carbon and Other Biogeochemical Cycles," in *Climate Change 2013,* figure 6.1, 471.

32 **warming, which will release even more carbon:** T. W. Crowther et
al., "Quantifying Global Soil Carbon Losses in Response to Warming," *Nature* 540, no. 7631 (2016): 104–8, https://doi.org/10.1038/nature20150.

32 **basically screamed from the pages of *Nature*:** Timothy M. Lenton
et al., "Climate Tipping Points—Too Risky to Bet Against," *Nature* 575
(2019): 592–95, https://www.nature.com/articles/d41586-019-03595-0#
correction-0.

32 **Sea level has already risen:** C. C. Hay et al., "Probabilistic Reanalysis
of Twentieth-Century Sea-Level Rise," *Nature* 517 (2015): 481–84; S. Dan-
gendorf et al., "Reassessment of 20th Century Global Mean Sea Level Rise,"
*Proceedings of the National Academy of Sciences of the United States of Amer-
ica* 114 (2017): 5946–51.

32 **higher and more frequent coastal flooding:** S. A. Kulp and B. H.
Strauss, "New Elevation Data Triple Estimates of Global Vulnerability to
Sea-Level Rise and Coastal Flooding," *Nature Communications* 10, no. 1
(2019): 4844, https://doi.org/10.1038/s41467-019-12808-z.

32 **Parts of New Orleans are already being abandoned:** Sophie Ka-
sakove, "Confronting Climate Change, Louisiana Shifts Toward Retreat,"

Pacific Standard, June 13, 2019, https://psmag.com/environment/confronting
-climate-change-louisiana-shifts-from-resilience-to-retreat.

32   **Miami is not far behind:** B. H. Strauss, S. Kulp, and A. Levermann, "Carbon Choices Determine US Cities Committed to Futures Below Sea Level," *Proceedings of the National Academy of Sciences of the United States of America* 112, no. 44 (2015): 13508–13, figure 3, https://doi.org/10.1073/pnas.1511186112.

32–33   **further twenty- to thirty-centimeter rise by 2050:** Kulp and Strauss, "New Elevation Data," 4844.

33   **sea level rise will continue beyond 2100:** Stocker et al., "Technical Summary," 105.

33   **large-scale migration away from unprotected coastlines:** Kulp and Strauss, "New Elevation Data," 4844.

33   **online sea level rise map:** NOAA Sea Level Rise Viewer, June 29, 2020, https://coast.noaa.gov/slr/.

33   **San Francisco is planning for one and a half feet:** NOAA Sea Level Rise Viewer: San Francisco in Year 2050, https://coast.noaa.gov/slr/#/layer/sce/8/-13628362.439392578/4548774.6611104105/14/satellite/86/0.8/2050/high/midAccretion.

33   **New York City expects two feet:** NOAA Sea Level Rise Viewer: New York, New York in Year 2050, https://coast.noaa.gov/slr/#/layer/sce/0/-8249569.088455777/4972872.553040696/12/satellite/87/0.8/2050/high/midAccretion.

Chapter 2

38   **no more big abalone shells:** L. Rogers-Bennett and C. A. Catton, "Marine Heat Wave and Multiple Stressors Tip Bull Kelp Forest to Sea Urchin Barrens," *Scientific Reports* 9, 15050 (2019), https://doi.org/10.1038/s41598-019-51114-y; Richie Hertzberg, "California's Disappearing Sea Snails Carry a Grim Climate Warning," *National Geographic,* August 20, 2019, https://www.nationalgeographic.com/environment/2019/08/red-abalone-closure-kelp-die-off-documentary-environment/.

39   **two-thirds of the fruits and nuts:** California Department of Food and Agriculture, *California Agricultural Statistics Review, 2018–2019* (2019), 1–58.

40   **perceives the degraded condition:** Peter H. Kahn Jr. and Thea Weiss, "The Importance of Children Interacting with Big Nature," *Children, Youth and Environments* 27, no. 2 (2017): 7–24.

40   *one thousand times* **faster:** Mark D. A. Rounsevell et al., "A Biodiversity Target Based on Species Extinctions," *Science* 368, no. 6496 (2020): 1193–95, https://doi.org/10.1126/science.aba6592.

40   **No one born after 1985:** Richard B. Rood, "Let's Call It: 30 Years of Above Average Temperatures Means the Climate Has Changed," The Conversation, February 26, 2015, https://theconversation.com/lets-call-it-30-years-of-above-average-temperatures-means-the-climate-has-changed-36175.

40   **just stop remarking on warmer weather:** Frances C. Moore et al., "Rapidly Declining Remarkability of Temperature Anomalies May Obscure Public Perception of Climate Change," *Proceedings of the National Academy of Sciences of the United States of America* 116, no. 11 (2019): 4905–10.

40   **still made people miserable:** UC Davis, "Tweets Tell Scientists How Quickly We Normalize Unusual Weather," Phys.org, February 25, 2019, https://phys.org/news/2019-02-tweets-scientists-quickly-unusual-weather.html.

41   **Eventually all life on Earth:** Alexander James Mustill, "The Fate of the Solar System" (public lecture, Kulturnatten, Lund University, September 21, 2019). See also Alexander James Mustill, "How to Kill a Planet," http://www.astro.lu.se/~alex/#PostMS.

41   **93 billion light-years:** Chris Baraniuk, "It Took Centuries, but We Now Know the Size of the Universe," *BBC Earth,* June 13, 2016, http://www.bbc.com/earth/story/20160610-it-took-centuries-but-we-now-know-the-size-of-the-universe.

42   **82 percent of biomass:** For all biomass data in this paragraph and the next, see Yinon M. Bar-On, Rob Phillips, and Ron Milo, "The Biomass Distribution on Earth," *Proceedings of the National Academy of Sciences of the United States of America* 115, no. 25 (2018): Table 1, https://doi.org/10.1073/pnas.1711842115.

42   **mollusks:** Paul Bunje, "The Mollusca: Sea Slugs, Squid, Snails, and Scallops," University of California Museum of Paleontology (2003), https://ucmp.berkeley.edu/taxa/inverts/mollusca/mollusca.php.

43   **five (!) previous mass extinctions:** Michael Greshko, "What Are Mass Extinctions, and What Causes Them?" *National Geographic,* September 26, 2019, https://www.nationalgeographic.com/science/prehistoric-world/mass-extinction/.

43   **humankind's most enduring legacy:** Elizabeth Kolbert, *The Sixth Extinction: An Unnatural History* (New York: Macmillan USA, 2014), 269.

43   **blue-green algae . . . then land plants:** Stephen Porder, "World Changers 3.0," *Natural History,* November 1, 2014, https://www.naturalhistorymag.com/features/232772/world-changers-30.

43   **Quaternary megafauna extinction:** Christopher Sandom et al., "Global Late Quaternary Megafauna Extinctions Linked to Humans, Not Climate Change," *Proceedings of the Royal Society B-Biological Sciences* 281, no. 1787 (2014), https://doi.org/10.1098/rspb.2013.3254.

44   **where mammals used to live:** S. Faurby, J. C. Svenning, and George Stevens, "Historic and Prehistoric Human-Driven Extinctions Have Reshaped Global Mammal Diversity Patterns," *Diversity and Distributions* 21, no. 10 (2015): 1155–66, https://doi.org/10.1111/ddi.12369.

44   **potentially aided by environmental changes:** Michael Balter, "What Killed the Great Beasts of North America?" *Science,* January 28, 2014, https://www.sciencemag.org/news/2014/01/what-killed-great-beasts-north-america; Lisa Nagaoka, Torben Rick, and Steven Wolverton, "The Overkill Model and Its Impact on Environmental Research," *Ecology and Evolution* 8, no. 19 (2018): 9683–96, https://doi.org/10.1002/ece3.4393.

44   **armadillo-like glyptodon:** Laura Geggel, "10 Extinct Giants That Once Roamed North America," Live Science, August 15, 2015, https://www.livescience.com/51793-extinct-ice-age-megafauna.html.

44   **elephant-like mammoths ranged:** Neotoma Paleoecology Database, age range: 15,000 years before present; taxon: Mammut americanum, http://wnapps.neotomadb.org/explorer/.

44   **hotspots for infectious disease:** Christopher E. Doughty et al., "Megafauna Decline Have Reduced Pathogen Dispersal Which May Have Increased Emergent Infectious Diseases," *Ecography* 43, no. 8 (2020): 1107–17, https://doi.org/10.1111/ecog.05209.

44 **come into contact with new animal species:** John Vidal, "Destruction of Habitat and Loss of Biodiversity Are Creating the Perfect Conditions for Diseases like COVID-19 to Emerge," Ensia, March 17, 2020, https://ensia.com/features/covid-19-coronavirus-biodiversity-planetary-health-zoonoses/.

45 **Sixty percent of emerging infectious diseases:** Kate E. Jones et al., "Global Trends in Emerging Infectious Diseases," *Nature* 451 (2008): 990–93, https://www.nature.com/articles/nature06536/.

45 **Most recent pandemics in people:** S. S. Morse et al., "Prediction and Prevention of the Next Pandemic Zoonosis," *Lancet* 380 (2012): 1956–65.

45 **half of large land mammal species:** Bar-On et al., "Biomass Distribution on Earth," 3.

45 **hefty Labrador:** The reference point in Bar-On et al., "Biomass Distribution on Earth," is over forty kilograms, or eighty-eight pounds. The American Kennel Club says male Labrador retrievers weigh between sixty-five and eighty pounds, but my family has never had a Lab that skinny. American Kennel Club, "Breed Weight Chart," https://www.akc.org/expert-advice/nutrition/breed-weight-chart/.

45 **naturalist John James Audubon:** John James Audubon, *Ornithological Biography* (1831), https://en.wikisource.org/wiki/Ornithological_Biography/Volume_1/Passenger_Pigeon.

45 **very last passenger pigeon:** Barry Yeoman, "Why the Passenger Pigeon Went Extinct," *Audubon,* May–June 2014, https://www.audubon.org/magazine/may-june-2014/why-passenger-pigeon-went-extinct.

46 **half as alive as it used to be:** Bar-On et al., "Biomass Distribution on Earth," 3.

46 **40 percent of forests:** Will Steffen et al., "Planetary Boundaries: Guiding Human Development on a Changing Planet," *Science* 347, no. 6223 (2015): figure S6, https://doi.org/10.1126/science.1259855.

46 **"human colonization of global ecosystems":** Helmut Haberl, Karl-Heinz Erb, and Fridolin Krausmann, "Human Appropriation of Net Primary Production: Patterns, Trends, and Planetary Boundaries," *Annual Review of Environment and Resources* 39 (2014): 367, 384, https://doi.org/10.1146/annurev-environ-121912-094620.

46 **15 percent of the original biomass:** Bar-On et al., "Biomass Distribution on Earth," 3, 37 (0.003/0.02 Gt).

46 **I am running out of adjectives:** Ed Yong, "Wait, Have We Really Wiped Out 60 Percent of Animals?" *The Atlantic,* October 31, 2018, https://www.theatlantic.com/science/archive/2018/10/have-we-really-killed-60-percent-animals-1970/574549/; World Wide Fund for Nature, *Living Planet Report 2020—Bending the Curve of Biodiversity Loss,* ed. R. E. A. Almond, M. Grooten, and T. Petersen (Gland, Switzerland: World Wide Fund for Nature, 2020).

46 **millions of years of evolution:** Matt Davis, Søren Faurby, and Jens-Christian Svenning, "Mammal Diversity Will Take Millions of Years to Recover from the Current Biodiversity Crisis," *Proceedings of the National Academy of Sciences of the United States of America* 115, no. 44 (2018): 11262–67, https://doi.org/10.1073/pnas.1804906115.

47 **barnyard animals now outweigh wild:** Bar-On et al., "Biomass Distribution on Earth," Table 1.

47 **Fishermen in Key West, Florida:** Loren McClenachan, "Documenting Loss of Large Trophy Fish from the Florida Keys with Historical Photo-

graphs," *Conservation Biology* 23, no. 3 (2009): 636–43, https://doi.org/10
.1111/j.1523-1739.2008.01152.x.

47 **"The Great Acceleration":** Will Steffen et al., "The Trajectory of the
Anthropocene: The Great Acceleration," *The Anthropocene Review* 2, no. 1
(2015): 81–98, https://doi.org/10.1177/2053019614564785.

47 **comprehensive 2019 global biodiversity assessment:** Intergovernmental Science-Policy Platform on Biodiversity and Ecosystem Services
(IPBES), *Summary for Policymakers, Global Assessment Report on Biodiversity and Ecosystem Services* (Bonn, Germany: IPBES Secretariat, 2019), 28,
figure 2.

47 **little scope to further expand fisheries:** Food and Agriculture Organization of the United Nations (FAO), *The State of World Fisheries and Aqua-
culture 2016: Contributing to Food Security and Nutrition for All* (Rome: FAO,
2016), 38.

47 **nature to be "deteriorating worldwide":** IPBES, *Summary for Policymakers, Global Assessment Report,* 10–11.

48 **"just slash up all the paintings":** Ken Caldeira, "Ocean Acidification:
Adaptive Challenge or Extinction Threat?" Special Lecture at American Geophysical Union Fall Meeting, San Francisco, California, December 6, 2012,
quote starts 51:37, https://www.youtube.com/watch?v=Pfz2l29aX9c.

48 **we're the asteroid:** Kevin Anderson was the first person I heard say humanity is like a conscious meteorite hitting the Earth. In December 2019, I saw the
public art installation *We Are the Asteroid III* by Justin Brice Guariglia.

49 **oxygen levels plummeted and crop yields declined:** Carl Zimmer,
"The Lost History of One of the World's Strangest Science Experiments," *The
New York Times,* March 29, 2019, https://www.nytimes.com/2019/03/29/sun
day-review/biosphere-2-climate-change.html.

49 **relatively stable climate:** K. D. Burke et al., "Pliocene and Eocene Provide Best Analogs for Near-Future Climates," *Proceedings of the National
Academy of Sciences of the United States of America* 115, no. 52 (2018):
13288–93, https://doi.org/10.1073/pnas.1809600115.

50 **build a relationship with nature:** University of Washington, College
of the Environment, "Urban Forestry / Urban Greening Research: Place
Attachment and Meaning," 2018, https://depts.washington.edu/hhwb/Thm
_Place.html.

50–51 **have lifelong impacts:** Nicole M. Ardoin, Alison W. Bowers, and Estelle
Gaillard, "Environmental Education Outcomes for Conservation: A Systematic Review," *Biological Conservation* 241 (January 2020): 108224, https://doi
.org/10.1016/j.biocon.2019.108224.

Chapter 3

53 **increasingly exceeding these planetary boundaries:** Johan Rockström et al., "A Safe Operating Space for Humanity," *Nature* 461, no. 7263
(2009): 472–75, http://dx.doi.org/10.1038/461472a; Will Steffen et al., "Planetary Boundaries: Guiding Human Development on a Changing Planet," *Science* 347, no. 6223 (2015), https://doi.org/10.1126/science.1259855.

54 **cutting down the very last tree:** Chris Field used this metaphor at
the "Our Common Future Under Climate Change" conference in Paris,
July 2015. Kimberly Nicholas (@KA_Nicholas), "We've never said, let's cut
the very last tree, catch the last fish in the ocean. Why burn last carbon? @

cfieldciwedu #IPCC #CFCC15," Twitter, July 10, 2015, 1:29 P.M., https://twitter.com/KA_Nicholas/status/619468426260676608?s.

54   **80 percent of land and 90 percent of the ocean:** James E. Watson et al., "Protect the Last of the Wild," *Nature* 563 (2018): 27–30, http://doi.org/10.1038/d41586-018-07183-6.

55   **opportunity and abundance, not just scarcity:** Jonathan Foley, "Living by the Lessons of the Planet," *Science* 356, no. 6335 (2017): 251–52, https://doi.org/10.1126/science.aal4863.

56   **"safe and just space for humanity":** Kate Raworth, "Safe and Just Space for Humanity: Can We Live Within the Doughnut?" (Oxfam Discussion Paper, February 2012), https://www.oxfam.org/en/research/safe-and-just-space-humanity; Julia Steinberger, Living Well Within Limits, University of Leeds, https://lili.leeds.ac.uk/.

56   **"change radically and for the better":** *Merriam-Webster,* s.v. "regenerate," https://www.merriam-webster.com/dictionary/regenerating.

58   **health, diversity, redundancy, and connectivity:** Carl Folke et al., "Social-Ecological Resilience and Biosphere-Based Sustainability Science," *Ecology and Society* 21, no. 3 (2016), https://doi.org/10.5751/es-08748-210341.

59   **Protecting remaining intact habitats:** B. W. Griscom et al., "Natural Climate Solutions," *Proceedings of the National Academy of Sciences of the United States of America* 114, no. 44 (2017): 11645–50, https://doi.org/10.1073/pnas.1710465114.

59   **we can't count on sick, stressed ecosystems:** W. R. L. Anderegg et al., "Climate-Driven Risks to the Climate Mitigation Potential of Forests," *Science* 368, no. 6497 (2020), https://doi.org/10.1126/science.aaz7005.

59   **planting trees . . . cannot undo the carbon pollution:** Michael Marshall, "Planting Trees Doesn't Always Help with Climate Change," *BBC Future,* May 26, 2020, https://www.bbc.com/future/article/20200521-planting-trees-doesnt-always-help-with-climate-change.

59   **makes sense for local conditions:** Bruno Locatelli et al., "Tropical Reforestation and Climate Change: Beyond Carbon," *Restoration Ecology* 23, no. 4 (2015): 337–43, https://doi.org/10.1111/rec.12209.

59   **separate strategies for reducing and *preventing*:** Duncan P. McLaren et al., "Beyond 'Net-Zero': A Case for Separate Targets for Emissions Reduction and Negative Emissions," *Frontiers in Climate* 1, no. 4 (2019), https://doi.org/10.3389/fclim.2019.00004.

60   **first start to see signs:** B. H. Samset, J. S. Fuglestvedt, and M. T. Lund, "Delayed Emergence of a Global Temperature Response after Emission Mitigation," *Nature Communications* 11, no. 1 (2020): figure 1, https://doi.org/10.1038/s41467-020-17001-1.

60   **dangerously close to deadly tipping points:** Timothy M. Lenton et al., "Climate Tipping Points—Too Risky to Bet Against," *Nature* 575 (2019): 592–95, https://www.nature.com/articles/d41586-019-03595-0#correction-0.

61   **The Stone Age didn't end because we ran out of stones:** Garson O'Toole, "'The Stone Age Did Not End Because the World Ran Out of Stones, and the Oil Age Will Not End Because We Run Out of Oil,'" Quote Investigator, January 7, 2018, https://quoteinvestigator.com/2018/01/07/stone-age/.

61   **attempts to put off the inevitable:** Intergovernmental Panel on Climate Change (IPCC), "Summary for Policymakers," in *Global Warming of*

1.5°C. An IPCC Special Report on the Impacts of Global Warming of 1.5°C Above Pre-Industrial Levels and Related Global Greenhouse Gas Emission Pathways, in the Context of Strengthening the Global Response to the Threat of Climate Change, Sustainable Development, and Efforts to Eradicate Poverty, ed. V. Masson-Delmotte et al. (Geneva, Switzerland: IPCC, 2018), figure SMP.3b. All pathways to 1.5°C shown involve the near elimination (91–97 percent reduction) of CO_2 emissions by 2050, with renewables generating 63–81 percent of electricity. The most sustainable and least risky pathways project a near majority (48–60 percent) of electricity coming from renewables globally by 2030. Even in a 2°C world, fossils are gone a few decades later.

61   **study published in *Nature*:** C. McGlade and P. Ekins, "The Geographical Distribution of Fossil Fuels Unused When Limiting Global Warming to 2 Degrees C," *Nature* 517, no. 7533 (2015): 187–90, https://doi.org/10.1038/nature14016.

61   **governments . . . and fossil fuel companies:** SEI, IISD, ODI, Climate Analytics, CICERO, and UNEP, "The Production Gap: The Discrepancy Between Countries' Planned Fossil Fuel Production and Global Production Levels Consistent with Limiting Warming to 1.5°C or 2°C," 2019, http://productiongap.org/.

62   **start shutting down existing fossil infrastructure early:** D. Tong et al., "Committed Emissions from Existing Energy Infrastructure Jeopardize 1.5 Degrees C Climate Target," *Nature* 572, no. 7769 (2019): 373–77, https://doi.org/10.1038/s41586-019-1364-3.

62   **If we can cut:** IPCC, *Global Warming of 1.5°C,* statement C1, and p. 95–96; Damon H. Matthews et al., "Opportunities and Challenges in Using Remaining Carbon Budgets to Guide Climate Policy," *Nature Geoscience* 13, no. 12 (2020): figure 2, https://doi.org/10.1038/s41561-020-00663-3.

62   **deep, rapid emissions reductions in every sector:** IPCC, *Global Warming of 1.5°C,* statement C2.

63   **"relatively straightforward to decarbonize":** S. J. Davis et al., "Net-Zero Emissions Energy Systems," *Science* 360, no. 6396 (June 29, 2018): 1, 7, figure 2.

63   **no shortage of creative and effective solutions:** For a roundup of solutions, see Kimberly Nicholas, "Climate Change 'We Can Fix It World Café,'" Kimnicholas.com, http://www.kimnicholas.com/we-can-fix-it-world-cafe.html.

63   **80 percent of global energy:** Hannah Ritchie and Max Roser, "Energy," Our World in Data, 2014, https://ourworldindata.org/energy.

63   **emissions from today's three biggest climate polluters:** Jeff Tollefson, "The Hard Truths of Climate Change—by the Numbers," *Nature* 573 (2019): 325–27, https://www.nature.com/immersive/d41586-019-02711-4/index.html.

63   **fossil CO_2 emissions 45 percent:** Global Carbon Project, "Global Carbon Budget 2019" (PowerPoint presentation, 2019), slide 30, https://www.globalcarbonproject.org/carbonbudget/19/presentation.htm.

63   **would take 363 years:** James Temple, "At This Rate, It's Going to Take Nearly 400 Years to Transform the Energy System," *MIT Technology Review,* March 14, 2018, https://www.technologyreview.com/2018/03/14/67154/at-this-rate-its-going-to-take-nearly-400-years-to-transform-the-energy-system/.

Chapter 4

68   **One grower I interviewed:** Kimberly Nicholas, "Will We Still Enjoy
Pinot Noir?" *Scientific American,* January 2016, 60–67; Kimberly Nicholas
Cahill, "Global Changes in Local Places: Climate Change and the Future of
the Wine Industry in Sonoma and Napa, California," PhD diss., Stanford
University, 2009, 94, https://pqdtopen.proquest.com/doc/305008154.html
?FMT=ABS.

69   **"sliding off to certain death":** "Kebnekaise," Summitpost.org, accessed
October 22, 2020, https://www.summitpost.org/kebnekaise/151165.

69   **official height of the mountain:** Brigit Katz, "Climate Change Has
Shrunk Sweden's Highest Peak," *Smithsonian,* September 10, 2019, https://
www.smithsonianmag.com/smart-news/climate-change-has-shrunk-swedens
-highest-peak-180973092/.

69   **oldest living creatures on Earth:** Robert Hudson Westover, "Methuse-
lah, a Bristlecone Pine Is Thought to Be the Oldest Living Organism on
Earth," United States Department of Agriculture, February 21, 2017, https://
www.usda.gov/media/blog/2011/04/21/methuselah-bristlecone-pine-thought-be
-oldest-living-organism-earth.

70   **policy relevant and yet policy-neutral:** Intergovernmental Panel on
Climate Change (IPCC), "Organization," n.d., https://archive.ipcc.ch/organi
zation/organization.shtml.

71   **wishes of 60 percent of Americans:** Cary Funk, "Key Findings About
Americans' Confidence in Science and Their Views on Scientists' Role in
Society," Pew Research Center, February 12, 2020, https://www.pewre
search.org/fact-tank/2020/02/12/key-findings-about-americans-confidence-in
-science-and-their-views-on-scientists-role-in-society/.

71   **feelings, manifested as physical sensations:** Lauri Nummenmaa et
al., "Bodily Maps of Emotions," *Proceedings of the National Academy of Sciences of the United States of America* 111, no. 2 (2014): 646–51, http://doi.org
/10.1073/pnas.1321664111.

71   **communities in Bangladesh:** Simon Ingram, "A Gathering Storm: Climate Change Clouds the Future of Children in Bangladesh," UNICEF Bangladesh, March 2019, https://www.unicef.org/rosa/reports/gathering-storm.

72   **3 billion animals:** "Australia's Fires 'Killed or Harmed Three Billion Animals,'" BBC News, July 28, 2020, https://www.bbc.com/news/world-australia
-53549936.

72   **heart that simply breaks:** Katherine Wilkinson, "How Empowering
Women and Girls Can Help Stop Global Warming," TEDWomen 2018, November 2018, video, 13:40, at 02:39, https://www.ted.com/talks/kathar
ine_wilkinson_how_empowering_women_and_girls_can_help_stop_global
_warming?language=en.

72   **broken-open heart:** Maddie Bender, "Climate Change Is Heartbreaking.
We Can Turn That Pain Towards Action," Massive Science, September 26,
2019, https://massivesci.com/articles/katharine-wilkinson-maddie-bender
-climate-change-heartbreak/.

72   **"stewards of grief":** Leehi Yona (@LeehiYona), "I think our role, above
all, is to be stewards of grief, to hold the hand of society as we enter the unknown space of the climate crisis. As scientists, we've dealt with our grief

more than others might have (though the ways scientists deny their grief is a diff lengthy post). 9/," Twitter, October 8, 2018, 10:21 P.M., https://twitter .com/LeehiYona/status/1049394285047558144?s.

74 **a warming lake ... bark beetle attacks:** Carolyn Wilke, "Lake Tahoe's Waters Continue to Warm, and Thousands of Trees Are Dying," *The Sacramento Bee*, July 27, 2017, https://www.sacbee.com/news/local/environment /article163973842.html; Jason Daley, "California's Drought Killed Almost 150 Million Trees," *Smithsonian*, July 10, 2019, https://www.smithsonianmag .com/smart-news/why-californias-drought-killed-almost-150-million-trees-1809 72591/.

76 **fresh air to see May Boeve:** Kimberly Nicholas (@KA_Nicholas), "Be honest: a lot has already been lost. We have to keep #fossilfuels in the ground. Need hope, grounded in reality @mayboeve. #ClimateAction," Twitter, May 6, 2016, 9:59 P.M., https://twitter.com/KA_Nicholas/status/72867556135 5259908.

77 **death of the last male northern white rhinoceros:** Sarah Gibbens, "After Last Male's Death, Is the Northern White Rhino Doomed?" *National Geographic*, March 20, 2018, https://www.nationalgeographic.com/news /2018/03/northern-white-rhino-male-sudan-death-extinction-spd/.

77 **relocate its capital city:** Jonathan Watts, "Indonesia Announces Site of Capital City to Replace Sinking Jakarta," *The Guardian*, August 26, 2019, https://www.theguardian.com/world/2019/aug/26/indonesia-new-capital-city -borneo-forests-jakarta.

77 **Louisiana has a new plan:** Christopher Flavelle and Mira Rojanasakul, "Louisiana Unveils Ambitious Plan to Help People Get Out of the Way of Climate Change," Bloomberg, May 15, 2019, https://www.bloomberg.com /graphics/2019-louisiana-strategic-plan/.

78 **"that's just hubris":** Ayana Elizabeth Johnson, "Fish out of Hot Water," *Mothers of Invention* podcast, https://www.mothersofinvention.online/fishout ofhotwater.

78 **91 percent of the Great Barrier Reef bleached:** T. P. Hughes et al., "Global Warming and Recurrent Mass Bleaching of Corals," *Nature* 543, no. 7645 (2017): 373, https://doi.org/10.1038/nature21707.

78 **diving through tears:** Katie Langin, "Why Some Climate Scientists Are Saying No to Flying," *Science*, May 13, 2019, https://www.sciencemag.org /careers/2019/05/why-some-climate-scientists-are-saying-no-flying.

78 **following a progression of photographs:** O. Hoegh-Guldberg et al., "Coral Reefs Under Rapid Climate Change and Ocean Acidification," *Science* 318, no. 5857 (2007): 1737–42, https://doi.org/10.1126/science.1152509.

79 **half the shallow-water corals there have died:** Timothy M. Lenton et al., "Climate Tipping Points—Too Risky to Bet Against," *Nature* 575 (2019): 592–95, https://www.nature.com/articles/d41586-019-03595-0#cor rection-0.

81 **first funeral for a glacier:** "Iceland Holds Funeral for First Glacier Lost to Climate Change," *The Guardian*, August 19, 2019, https://www.theguardian .com/world/2019/aug/19/iceland-holds-funeral-for-first-glacier-lost-to-climate -change.

Chapter 5

85   **Long-term happiness depends on close relationships:** Liz Mineo, "Good Genes Are Nice, but Joy Is Better," *The Harvard Gazette,* April 11, 2017, https://news.harvard.edu/gazette/story/2017/04/over-nearly-80-years -harvard-study-has-been-showing-how-to-live-a-healthy-and-happy-life/.

85   **research shows that happiness:** Emily Esfahani Smith, "There's More to Life than Being Happy," *The Atlantic,* January 9, 2013, https://www .theatlantic.com/health/archive/2013/01/theres-more-to-life-than-being-happy /266805/.

85   **study of 12 million blogs:** C. Mogilner, S. D. Kamvar, and J. Aaker, "The Shifting Meaning of Happiness," *Social Psychological and Personality Science* 2, no. 4 (2011): 395–402.

85   **values behind meaning:** Esfahani Smith, "There's More to Life."

85   **"all joy and no fun":** Jennifer Senior, *All Joy and No Fun: The Paradox of Modern Parenthood* (New York: Ecco/HarperCollins, 2011).

85   **taking care of their children:** Daniel Gilbert, "Does Fatherhood Make You Happy?" *Time,* June 11, 2006, http://www.danielgilbert.com/Does %20Fatherhood%20Make%20You%20Happy.htm.

86   **Frankl describes three routes to meaning:** Viktor E. Frankl, *Man's Search for Meaning* (Boston: Beacon Press, 1959).

86   **life isn't about avoiding suffering:** Paul Kalanithi, *When Breath Becomes Air* (New York: Random House, 2016), 150.

87   **a role in the greatest story:** This framing draws from Login George and Crystal Park, who propose that meaning in life can be understood as consisting of three parts: experiencing one's life as making sense and being understandable as part of a coherent narrative (what they call comprehension); being directed and motivated by valued goals (purpose); and feeling one's existence is of significance, importance, relevance, and value in the world (mattering). Login S. George and Crystal L. Park, "Meaning in Life as Comprehension, Purpose, and Mattering: Toward Integration and New Research Questions," *Review of General Psychology* 20, no. 3 (2016): 205–20, https:// doi.org/10.1037/gpr0000077.

87   **a quarter of Americans:** R. Kobau et al., "Well-Being Assessment: An Evaluation of Well-Being Scales for Public Health and Population Estimates of Well-Being Among US Adults," *Applied Psychology: Health and Well-Being* 2 (2010): 272–97, doi:10.1111/j.1758-0854.2010.01035.x.

89   **win-win leisure choices:** Angela Druckman and Birgitta Gatersleben, "A Time-Use Approach: High Subjective Wellbeing, Low Carbon Leisure," *Journal of Public Mental Health* 18, no. 2 (2019): 85–93, https://doi.org/10.1108 /jpmh-04-2018-0024.

92   **Japanese concept of *ikigai*:** Yukari Mitsuhashi, "Ikigai: A Japanese Concept to Improve Work and Life," BBC Worklife, August 8, 2017, https://www .bbc.com/worklife/article/20170807-ikigai-a-japanese-concept-to-improve-work -and-life.

92   **Ira Glass, for your reassuring advice:** Daniel Sax, "THE GAP by Ira Glass," video, 2:18, https://vimeo.com/85040589.

93   **France pledging only to work:** "Taking Action for an Ecological Awakening!" and "How to Choose Your Job: The Right Questions to Ask," Pour un réveil écologique, https://pour-un-reveil-ecologique.org/en/.

93 **"to help shift society":** "We Are a Network of Advertising Insiders Working Together to Reshape Our Industry to Tackle Climate Change," Purpose Disruptors, accessed July 23, 2020, https://www.purposedisruptors.org/.

93 **in countries with high emissions rates:** This text draws from my personal essay published alongside our scientific study: Kimberly Nicholas, "A Hard Look in the Climate Mirror," *Scientific American,* July 12, 2017, https://blogs.scientificamerican.com/observations/a-hard-look-in-the-climate -mirror/.

93 **whether and how many children you want:** Natalia Kanem, "Family Planning Is a Human Right," United Nations Population Fund, July 6, 2018, accessed July 24, 2020, https://www.unfpa.org/press/family-planning -human-right.

94 **part of the "culture of death":** Courtney Kirchoff, "Environmentalists Openly Blaming People with 'Too Many Children' for Climate Change," *Louder with Crowder,* July 12, 2017, https://www.louderwithcrowder.com /americans-one-fewer-child-climate-change.

94 **"liberal-environmentalist suicide pact":** Rachel Lu, "The Problem with the 'Science' Behind Having Fewer Children for the Planet's Sake," *National Review,* July 15, 2017, https://www.nationalreview.com/2017/07/climate -change-study-population-reduction-childlessness-recommendation-preposter ous-carbon-footprint/.

94 **climate opinion-haver Roy Scranton:** Roy Scranton, "Raising My Child in a Doomed World," *The New York Times,* July 16, 2018, https://www .nytimes.com/2018/07/16/opinion/climate-change-parenting.html.

95 **vote of hope:** Victoria Whitley-Berry, "Want to Slow Global Warming? Researchers Look to Family Planning," National Public Radio, July 18, 2017, https://www.npr.org/2017/07/19/537954372/want-to-slow-global-warming-re searchers-look-to-family-planning?t=1584613475934&t=1595607097011.

95 **household emissions rising after a new baby:** J. Nordström, J. F. Shogren, and L. Thunström, "Do Parents Counter-balance the Carbon Emissions of Their Children?" *PLoS One* 15, no. 4 (2020): e0231105, https://doi .org/10.1371/journal.pone.0231105.

95 **reduced their household climate footprint:** Beth Buczynski, "How to Raise a Zero Carbon Footprint Baby: Author Keya Chatterjee," EcoSalon, August 4, 2013, http://ecosalon.com/raise-zero-carbon-footprint-baby/; "Raising a Zero Footprint Baby: Interview with author Keya Chatterjee," Earth-Share, https://www.earthshare.org/baby/.

96 **"central goals in life is to procreate":** Umair Irfan, "We Need to Talk About the Ethics of Having Children in a Warming World," Vox, March 11, 2019, https://www.vox.com/2019/3/11/18256166/climate-change-having -kids.

96 **"Do you want to be a mother?":** Michelle Kovacevic, "The Choice to Have Children in a Climate Crisis," Dumbo Feather, February 25, 2020, https://www .dumbofeather.com/articles/the-choice-to-have-children-in-a-climate-crisis/.

Chapter 6

97 **humans *fight the sun*:** National Research Center, *Climate Intervention: Reflecting Sunlight to Cool Earth* (Washington, DC: National Academies Press, 2015), https://doi.org/10.17226/18988.

98   **oceans to keep acidifying toward death:** Alan Robock, "20 Reasons Why Geoengineering May Be a Bad Idea," *Bulletin of the Atomic Scientists,* no. 2 (2008): 14–18, https://doi.org/10.2968/064002006.

98   **the sky would no longer be blue:** Michael D. Lemonick, "Geoengineered Sky: Bye-Bye Blue, Hello White," Climate Central, June 4, 2012, https://www.climatecentral.org/news/geoengineered-sky-bye-bye-blue-hello-white.

98   **never see the stars again:** Maddie Stone, "Geoengineering Could Be a Disaster for Astronomy," Gizmodo, December 28, 2016, https://gizmodo.com /geoengineering-could-be-a-disaster-for-astronomy-1790555051.

99   **lots of studies showing it's technically possible:** There is *so* much science on fixing climate change! A selection: Intergovernmental Panel on Climate Change (IPCC), "Summary for Policymakers," in *Climate Change 2014: Mitigation of Climate Change. Contribution of Working Group III to the Fifth Assessment Report of the Intergovernmental Panel on Climate Change,* ed. O. Edenhofer et al. (Cambridge, UK: Cambridge University Press, 2014); Project Drawdown, "The Drawdown Review," March 10, 2020, https://draw down.org/drawdown-review; J. Falk et al., "Exponential Roadmap 1.5: Scaling 36 Solutions to Halve Emissions by 2030," Future Earth, Sweden, September 2019, https://exponentialroadmap.org/; Kimberly Henderson et al., "Climate Math: What a 1.5 Degree Pathway Would Take," *McKinsey Quarterly,* April 30, 2020, https://www.mckinsey.com/business-functions/sustainability/our -insights/climate-math-what-a-1-point-5-degree-pathway-would-take#.

99   **"possible within the laws of chemistry and physics":** IPCC, "Summary for Policymakers of IPCC Special Report on Global Warming of 1.5°C Approved by Governments," October 8, 2018, https://www.ipcc.ch/2018/10 /08/summary-for-policymakers-of-ipcc-special-report-on-global-warming-of-1 -5c-approved-by-governments/.

100   **piece on climate activism by Jim Shultz:** Jim Shultz, "To My Friend the Climate Defeatist: Here's Why I'm Still in the Fight," *Yes!,* April 30, 2014, https://www.yesmagazine.org/environment/2014/04/30/to-my-friend-the-climate -defeatist-here-s-why-i-m-still-in-the-fight/.

101   **"Once we start to act":** Greta Thunberg, "The Disarming Case to Act Right Now on Climate Change," TEDx Stockholm, November 2018, transcript, https://www.ted.com/talks/greta_thunberg_the_disarming_case_to _act_right_now_on_climate_change/transcript.

101   **Rebecca Solnit's description of hope:** Rebecca Solnit, *Hope in the Dark: Untold Histories, Wild Possibilities* (New York: Nation Books, 2004), chapter 1.

101   **"a slope we slide down":** Kate Marvel, "Thinking About Climate on a Dark, Dismal Morning," *Scientific American,* December 25, 2018, https:// blogs.scientificamerican.com/hot-planet/thinking-about-climate-on-a-dark-dis mal-morning/.

101   **As the Heath brothers write:** Chip Heath and Dan Heath, *Switch: How to Change Things When Change Is Hard* (Waterville, ME: Thorndike Press, 2011), 153–56.

102   **colleagues from psychology:** These five stages are my own, though I have been inspired by the work of Per Espen Stoknes, Frida Hylander, and Kirsti Jylhä in formulating them.

103   **bizarre 1990 Earth Day special:** Aiz Ayob, "The Earth Day Special," n.d.,
YouTube video, 1:39:35, https://www.youtube.com/watch?v=Uz0HzS1O-ug.
Carl Sagan starts around 16:00.

103   **only half hear about it once a month:** Anthony Leiserowitz et al.,
"Climate Change in the American Mind: April 2019," Yale Program on Climate Change Communication, June 27, 2019, https://climatecommunication
.yale.edu/publications/climate-change-in-the-american-mind-april-2019/6/.

103   **media, governments, and schools rarely talk about climate:** Seth
Wynes and Kimberly Nicholas, "The Climate Mitigation Gap: Education and
Government Recommendations Miss the Most Effective Individual Actions,"
Environmental Research Letters 12, no. 7 (2017), table 1, https://doi.org/10
.1088/1748-9326/aa7541.

104   **conversations more than once a month:** Leiserowitz et al., "Climate
Change in the American Mind."

105   **from a safe, cozy house:** Andy Puddicombe, Mindfulness Meditation
featured on Headspace.com.

105   **face-to-face relationships with our neighbors:** See the great practical steps in "Ask Umbra's 21-Day Apathy Detox: You Can't Build Community
Without, Well, Your Community," Grist, n.d., https://grist.org/guides/umbra
-apathy-detox/you-cant-build-community-without-well-your-community/.

106   **decentralized, self-organized, local networks:** Christine Nieves,
"Why Community Is Our Best Chance for Survival—a Lesson Post–Hurricane
Maria," TedMed Talk, 2018, video, 16:59, https://www.tedmed.com/talks
/show?id=731041.

106   **pursuing purpose is part of achieving meaning:** Login S. George
and Crystal L. Park, "Meaning in Life as Comprehension, Purpose, and Mattering: Toward Integration and New Research Questions," *Review of General
Psychology* 20, no. 3 (2016): 205–20, https://doi.org/10.1037/gpr0000077.

106   **40 percent of Americans:** Rosemarie Kobau et al., "Well-Being Assessment: An Evaluation of Well-Being Scales for Public Health and Population
Estimates of Well-Being Among US Adults," *IAAP* 2, no. 3 (2010): 272–97,
https://doi.org/10.1111/j.1758-0854.2010.01035.x.

Chapter 7

111   **think-tank economist:** Irwin Stelzer, "Asking the Right Question on Climate Change," IrwinStelzer.com, September 17, 2019, https://irwinstelzer.com
/2019/09/17/asking-the-right-question-on-climate-change/.

112   **thermometer is not liberal or conservative:** Jonathan Watts, "Interview: Katharine Hayhoe: 'A Thermometer Is Not Liberal or Conservative,"
The Guardian, January 6, 2019, https://www.theguardian.com/science/2019
/jan/06/katharine-hayhoe-interview-climate-change-scientist-crisis-hope.

112   **unlike facts, every citizen is entitled to their own:** "Everyone is
entitled to his own opinion, but not to his own facts," attributed to Daniel
Patrick Moynihan in "An American Original," *Vanity Fair,* October 6, 2010,
https://www.vanityfair.com/news/2010/11/moynihan-letters-201011.

113   **"sheer weight of consistent evidence":** "The 97% Consensus on Global
Warming," Skeptical Science, n.d., https://www.skepticalscience.com/global
-warming-scientific-consensus.htm.

113 **distant relative of Greta Thunberg:** Amy Goodman, "School Strike for Climate: Meet 15-Year-Old Activist Greta Thunberg, Who Inspired a Global Movement," Democracy Now! December 11, 2018, https://www.democra cynow.org/2018/12/11/meet_the_15_year_old_swedish.

113 **suggested that burning coal could cause global warming:** Svante Arrhenius, "On the Influence of Carbonic Acid in the Air upon the Tempera- ture of the Ground," *The London, Edinburgh, and Dublin Philosophical Maga- zine and Journal of Science* Series 5, no. 41 (April 1896): 251.

114 **same range as the most sophisticated one today:** S. C. Sherwood et al., "An Assessment of Earth's Climate Sensitivity Using Multiple Lines of Evidence," *Reviews of Geophysics* 58, no. 4 (December 2020), https://doi.org /10.1029/2019RG000678.

114 **Those activities are:** Intergovernmental Panel on Climate Change (IPCC), "Summary for Policymakers," in *Climate Change 2014: Mitigation of Climate Change. Contribution of Working Group III to the Fifth Assessment Report of the Intergovernmental Panel on Climate Change,* ed. O. Edenhofer et al. (Cambridge, UK: Cambridge University Press, 2014), figure SPM.2, 8.

114 **found 0 studies out of 928:** N. Oreskes, "The Scientific Consensus on Climate Change," *Science* 306 (2004): 1686.

114 **99.94 percent consensus:** J. L. Powell, "The Consensus on Anthropo- genic Global Warming Matters," *Bulletin of Science, Technology and Society* 36 (2016): 157–63.

114 **consensus of 97.1 percent:** J. Cook et al., "Quantifying the Consensus on Anthropogenic Global Warming in the Scientific Literature," *Environmen- tal Research Letters* 8, no. 2 (2013): 024024, http://doi.org/10.1088/1748-9326 /8/2/024024.

114 **study of the studies of studies:** John Cook et al., "Consensus on Con- sensus: A Synthesis of Consensus Estimates on Human-Caused Global Warming," *Environmental Research Letters* 11, no. 4 (2016), https://doi.org /10.1088/1748-9326/11/4/048002.

114 **the greater the relevant expertise:** W. R. Anderegg et al., "Expert Credibility in Climate Change," *Proceedings of the National Academy of Sci- ences of the United States of America* 107, no. 27 (2010): 12107–9, https://doi .org/10.1073/pnas.1003187107.

114 **"gateway belief":** S. L. van der Linden et al., "The Scientific Consensus on Climate Change as a Gateway Belief: Experimental Evidence," *PLoS One* 10, no. 2 (2015): e0118489, https://doi.org/10.1371/journal.pone.0118489; D. Ding et al., "Support for Climate Policy and Societal Action Are Linked to Perceptions About Scientific Agreement," *Nature Climate Change* 1 (2011): 462–66, https://doi.org/10.1038/nclimate1295.

114 **only about half of Americans:** David M. Romps and Jean P. Retzinger, "Climate News Articles Lack Basic Climate Science," *Environmental Re- search Communication* 1, no. 38 (2019): 1–10, https://doi.org/10.1088/2515 -7620/ab37dd.

114 **scientific consensus was the least-covered topic:** Seth Wynes and Kimberly Nicholas, "Climate Science Curricula in Canadian Secondary Schools Focus on Human Warming, Not Scientific Consensus, Impacts or Solutions," *PLoS One* 14, no. 7 (2019): e0218305, https://doi.org/10.1371/jour nal.pone.0218305.

115 **the annual meeting of 1965:** Benjamin Franta, "Early Oil Industry Knowledge of CO_2 and Global Warming," *Nature Climate Change* 8 (2018): 1024–25, https://doi.org/10.1038/s41558-018-0349-9.

115 **organized disinformation campaigns:** Robert J. Brulle, "Networks of Opposition: A Structural Analysis of U.S. Climate Change Countermovement Coalitions 1989–2015," *Sociological Inquiry* (2019), https://doi.org/10.1111/soin.12333.

115 **fossil fuel emissions have tripled:** Global Carbon Project, "Global Carbon Budget 2019" (PowerPoint presentation, 2019), slide 12, https://www.globalcarbonproject.org/carbonbudget/19/presentation.htm.

116 **"ExxonMobil misled the public":** Geoffrey Supran and Naomi Oreskes, "Assessing ExxonMobil's Climate Change Communications (1977–2014)," *Environmental Research Letters* 12, no. 8 (2017): 1–19, https://doi.org/10.1088/1748-9326/aa815f.

116 **InfluenceMap analysis:** "Big Oil's Real Agenda on Climate Change," InfluenceMap, March 2019, https://influencemap.org/report/How-Big-Oil-Continues-to-Oppose-the-Paris-Agreement-38212275958aa21196dae3b76220bddc.

116 **they don't meet even the 2°C goal:** "Absolute Impact: Why Oil Majors' Climate Ambitions Fall Short of Paris Limits," Carbon Tracker, June 24, 2020, https://carbontracker.org/reports/absolute-impact/.

117 **29 percent . . . biofuels from algae:** "Big Oil's Real Agenda," InfluenceMap, March 2019, 3, 13.

117 **top one hundred oil companies are responsible for 71 percent:** Paul Griffin, "The Carbon Majors Database CDP Carbon Majors Report 2017," Carbon Disclosure Project, 2017.

117 **Lives and species are being lost:** J. Lelieveld et al., "Effects of Fossil Fuel and Total Anthropogenic Emission Removal on Public Health and Climate," *Proceedings of the National Academy of Sciences of the United States of America* 116, no. 15 (2019): 7192–97, https://doi.org/10.1073/pnas.1819989116; R. Licker et al., "Attributing Ocean Acidification to Major Carbon Producers," *Environmental Research Letters* 14, no. 12 (2019), https://doi.org/10.1088/1748-9326/ab5abc.

117 **"predatory delay":** Alex Steffen, "Predatory Delay and the Rights of Future Generations," Medium, April 26, 2016, https://medium.com/@AlexSteffen/predatory-delay-and-the-rights-of-future-generations-69b06094a16.

117 **only one in five Americans:** Anthony Leiserowitz et al., *Climate Change in the American Mind: December 2018* (New Haven, CT: Yale Program on Climate Change Communication, 2018), 9, https://climatecommunication.yale.edu/wp-content/uploads/2019/01/Climate-Change-American-Mind-December-2018.pdf.

117 **analyzed a 2019 letter:** Timothy Osborn et al., "Letter Signed by '500 Scientists' Relies on Inaccurate Claims About Climate Science," *Climate Feedback,* September 23, 2019, https://climatefeedback.org/evaluation/letter-signed-by-500-scientists-relies-on-inaccurate-claims-about-climate-science/.

118 **As Upton Sinclair wrote:** Garson O'Toole, "It Is Difficult to Get a Man to Understand Something When His Salary Depends Upon His Not Understanding It," Quote Investigator, November 30, 2017, https://quoteinvestigator.com/2017/11/30/salary/.

118 **avoid using the phrase "climate change":** Oliver Milman, "US Federal Department Is Censoring Use of Term 'Climate Change,' Emails Reveal," *The Guardian*, August 7, 2017, https://www.theguardian.com/environment/2017/aug/07/usda-climate-change-language-censorship-emails.

118 **climate or fossil fuels or even "science":** Laignee Barren, "Here's What the EPA's Website Looks Like After a Year of Climate Change Censorship," *Time*, March 1, 2018, https://time.com/5075265/epa-website-climate-change-censorship/.

118 **"nuisance flooding" instead of sea level rise:** Elizabeth Kolbert, "Miami Is Flooding," *The New Yorker*, December 21 and 28, 2015.

118 **every other female scientist:** Scott Waldemar and Niina Heikkinen, "As Climate Scientists Speak Out, Sexist Attacks Are on the Rise," *Scientific American*, August 22, 2018, https://www.scientificamerican.com/article/as-climate-scientists-speak-out-sexist-attacks-are-on-the-rise/.

119 **study across fifty-six countries:** Matthew J. Hornsey et al., "Meta-analyses of the Determinants and Outcomes of Belief in Climate Change," *Nature Climate Change* 6 (2016): 622–26, https://doi.org/10.1038/nclimate2943.

119 **Unique among major political parties:** Jonathan Chait, "Why Are Republicans the Only Climate-Science-Denying Party in the World?" *New York*, September 27, 2015, https://nymag.com/intelligencer/2015/09/whys-gop-only-science-denying-party-on-earth.html; David Roberts, "The GOP Is the World's Only Major Climate-Denialist Party. But Why?" Vox, December 2, 2015. https://www.vox.com/2015/12/2/9836566/republican-climate-denial-why.

119 **increasing divisions within the party:** Justin Gillis, "The Republican Climate Closet," *The New York Times*, August 12, 2019, https://www.nytimes.com/2019/08/12/opinion/republicans-environment.html.

119 **stronger than gun control or abortion:** Oliver Milman, "Climate Crisis More Politically Polarizing than Abortion for US Voters, Study Finds," *The Guardian*, May 22, 2019, https://www.theguardian.com/us-news/2019/may/21/climate-crisis-more-politically-polarizing-than-abortion-for-us-voters-study-finds.

120 **California under two different scenarios:** Katharine Hayhoe et al., "Emissions Pathways, Climate Change, and Impacts on California," *Proceedings of the National Academy of Sciences of the United States of America* 101, no. 34 (2004): 12422–27.

120 **wine industry drives:** Cathy Bussewitz, "Study: Wine Adds More than $13 Billion to Sonoma County Economy," *The Press Democrat* (Santa Rosa, CA), January 9, 2014, https://legacy.pressdemocrat.com/business/1853641-181/study-wine-adds-more-than?sba=AAS; Stonebridge Research Group, "The Economic Impact of Napa County's Wine and Grapes," November 2012, 8, https://napavintners.com/community/docs/napa_economic_impact_2012.pdf.

120 **"'another piece of climate alarmism'":** Kimberly Nicholas (@KA_Nicholas), "Hi John, thanks for your interest! I'll email you a PDF—the link not immediately obvious while searching on my phone from the train. Clarification: the @nytimes prominently reprinted American Enterprise Institute 'expert' calling @Stanford @UCBerkeley et al scientists alarmists," Twitter, July 9, 2019, 4:29 P.M., https://twitter.com/KA_Nicholas/status/1148599946247098369?s.

121 **Koch-funded group with long ties to tobacco:** "Competitive Enterprise Institute," SourceWatch, accessed July 25, 2020, https://www.source watch.org/index.php/Competitive_Enterprise_Institute.

121 **led opposition to the Paris Agreement:** Lisa Friedman and Hiroko Tabuchi, "Following the Money That Undermines Climate Science," *The New York Times,* July 10, 2019, https://www.nytimes.com/2019/07/10/climate /nyt-climate-newsletter-cei.html.

121 **twice as likely to cover statements opposing climate action:** R. Wetts, "In Climate News, Statements from Large Businesses and Opponents of Climate Action Receive Heightened Visibility," *Proceedings of the National Academy of Sciences of the United States of America* 117, no. 32 (2020): 19054–60, https://doi.org/10.1073/pnas.1921526117.

121 **"genuine tipping point or just symbolic":** Reddit Ask Me Anything, "AskScience AMA Series: We Are Climate Scientists Here to Talk About the Important Individual Choices You Can Make to Help Mitigate Climate Change. Ask Us Anything!" November 9, 2017, https://www.reddit.com/r/askscience /comments/7bsv2d/askscience_ama_series_we_are_climate_scientists/.

122 **"past experience may be a poor guide":** Kimberly Nicholas and William H. Durham, "Farm-Scale Adaptation and Vulnerability to Environmental Stresses: Insights from Winegrowing in Northern California," *Global Environmental Change—Human and Policy Dimensions* 22, no. 2 (May 2012): 491, https://doi.org/10.1016/j.gloenvcha.2012.01.001, referring to climate analysis in D. B. Lobell et al., "Impacts of Future Climate Change on California Perennial Crop Yields: Model Projections with Climate and Crop Uncertainties," *Agricultural and Forest Meteorology* 141 (2006): 208–18.

123 **"decades ahead of even the most pessimistic climate models":** Fiona Harvey, "Tropical Forests Losing Their Ability to Absorb Carbon, Study Finds," *The Guardian,* March 4, 2020, https://www.theguardian.com /environment/2020/mar/04/tropical-forests-losing-their-ability-to-absorb-car bon-study-finds.

124 **10 percent as "climate dismissives":** M. Goldberg et al., "For the First Time, the Alarmed Are Now the Largest of Global Warming's Six Americas," Yale Program on Climate Change Communication, January 16, 2020, https:// climatecommunication.yale.edu/publications/for-the-first-time-the-alarmed-are -now-the-largest-of-global-warmings-six-americas/.

124 **beliefs that justify and promote existing hierarchies:** Kirsti M. Jylhä and Nazar Akrami, "Social Dominance Orientation and Climate Change Denial: The Role of Dominance and System Justification," *Personality and Individual Differences* 86 (2015): 108–11, https://doi.org/10.1016 /j.paid.2015.05.041; Kirsti M. Jylhä and Kahl Hellmer, "Right-Wing Populism and Climate Change Denial: The Roles of Exclusionary and Anti-Egalitarian Preferences, Conservative Ideology, and Antiestablishment Attitudes," *Analyses of Social Issues and Public Policy* (2020), https://doi.org/10.1111/asap.12203.

124 **constructive conversations with the 90 percent:** Carol Peters, "Talking About Climate: Build Consensus Through Shared Values and Common Ground," Rutgers Institute of Earth, Ocean, and Atmospheric Sciences, March 6, 2019, https://eoas.rutgers.edu/talking-about-climate-build -consensus-through-shared-values-and-common-ground/.

124 **understanding and overcoming climate change denialism:** Gabrielle Wong-Parodi and Irina Feygina, "Understanding and Countering the

Motivated Roots of Climate Change Denial," *Current Opinion in Environmental Sustainability* 42 (2020): 60–64, https://doi.org/10.1016/j.cosust.2019.11.008.

125 **"maintain the potential to shift":** Hank C. Jenkins-Smith et al., "Partisan Asymmetry in Temporal Stability of Climate Change Beliefs," *Nature Climate Change* 10 (2020): 322–28, https://doi.org/10.1038/s41558-020-0719-y.

125 **depolarized respondent judgments:** Sander van der Linden, Anthony Leiserowitz, and Edward Maibach, "Scientific Agreement Can Neutralize Politicization of Facts," *Nature Human Behavior* 2 (2018): 2–3, https://doi.org/10.1038/s41562-017-0259-2.

125 **the American public overwhelmingly does trust scientists:** Cary Funk, "Key Findings About Americans' Confidence in Science and Their Views on Scientists' Role in Society," Pew Research Center, February 12, 2020, https://www.pewresearch.org/fact-tank/2020/02/12/key-findings-about-americans-confidence-in-science-and-their-views-on-scientists-role-in-society/.

125 **Yale "Climate Change in the American Mind" surveys:** Abel Gustafson et al., "A Growing Majority of Americans Think Global Warming Is Happening and Are Worried," Yale Center for Climate Change Communication, February 21, 2019, https://climatecommunication.yale.edu/publications/a-growing-majority-of-americans-think-global-warming-is-happening-and-are-worried/.

126 **"the impacts of climate change are already being felt":** USGCRP, *Impacts, Risks, and Adaptation in the United States: Fourth National Climate Assessment,* vol. 2 (Washington, DC: US Global Change Research Program, 2018), http://doi.org/10.7930/NCA4.2018.

126 **girls who do the same work as boys:** Tom Richell, "International Women's Day: Norwegian Child Social Experiment Brutally Illustrates Gender Inequality," *Independent,* March 9, 2018, https://www.independent.co.uk/news/world/international-womens-day-norway-children-video-gender-pay-gap-boys-girls-finansforbundet-trade-union-a8245841.html.

Chapter 8

131 **"dream of studying the ocean":** Kimberly Nicholas (@KA_Nicholas), "Dreams of studying the ocean- but afraid the ocean may be dead when she grows up. #peopleofclimate #PCM #Copenhagen," Twitter, September 21, 2014, 3:02 P.M., https://twitter.com/KA_Nicholas/status/513674582109478913.

132 **most benefit other sustainability goals:** D. Lusseau and F. Mancini, "Income-Based Variation in Sustainable Development Goal Interaction Networks," *Nature Sustainability* 2 (2019): 242–47.

133 **empowering and educating women:** W. Lutz, "Global Sustainable Development Priorities 500 y After Luther: *Sola schola et sanitate,*" *Proceedings of the National Academy of Sciences of the United States of America* 114, no. 27 (2017): 6904–13; W. Lutz, "How Population Growth Relates to Climate Change," *Proceedings of the National Academy of Sciences of the United States of America* 114, no. 46 (2017): 12103–5, https://doi.org/10.1073/pnas.1717178114.

133–34 **Eight years later . . . permanent exodus:** Ben Casselman, "Katrina Washed Away New Orleans's Black Middle Class," FiveThirtyEight, August 24, 2015, https://fivethirtyeight.com/features/katrina-washed-away-new-orleanss-black-middle-class/.

134 **"little sign of acting":** Climate Action Tracker, "Governments Still Show-ing Little Sign of Acting on Climate Crisis," December 2019, https://climate actiontracker.org/documents/698/CAT_2019-12-10_BriefingCOP25_Warming ProjectionsGlobalUpdate_Dec2019.pdf.

135 **emitted nearly half . . . 90 percent . . . 13 percent:** All statistics in the paragraph come from the Global Carbon Project, "Global Carbon Budget 2019" (PowerPoint presentation, 2019), slides 70, 71, 87 (note: slide 70 is in tons carbon, not CO_2; multiply by 3.7 to convert), https://www.globalcarbo nproject.org/carbonbudget/19/presentation.htm.

136 **study led by Daniel O'Neill:** Daniel W. O'Neill et al., "A Good Life for All Within Planetary Boundaries," *Nature Sustainability* 1, no. 2 (2018): 88–95, https://doi.org/10.1038/s41893-018-0021-4.

137 **like there are four planet Earths:** Charlotte McDonald, "How Many Earths Do We Need?" BBC News, June 16, 2015, https://www.bbc.com /news/magazine-33133712.

137 **Seventy-two percent:** Edgar G. Hertwich and Glen P. Peters, "Carbon Footprint of Nations: A Global Trade-Linked Analysis," *Environmental Sci-ence and Technology* 43 (2009): 6414–20.

138 **incomes among the top 10 percent:** Stuart Capstick et al., "Bridging the Gap—The Role of Equitable Low-Carbon Lifestyles," chapter 6 in *UNEP Emissions Gap Report 2020*, figure 6.1, https://www.unep.org/emissions-gap -report-2020.

138 **half of household carbon pollution:** T. Gore, "Extreme Carbon In-equality: Why the Paris Climate Deal Must Put the Poorest, Lowest Emitting and Most Vulnerable People First," Oxfam International, 2015, http://policy practice.oxfam.org.uk/publications/extreme-carbon-inequality-why-the-paris climate-deal-must-put-the-poorest-lowes-582545.

138 **cause just 10 percent:** Hertwich and Peters, "Carbon Footprint of Na-tions," 6414–20.

138 **an Oxfam report:** Richard King, "Carbon Emissions and Income Inequal-ity: Technical Note," Oxfam International, 2015, https://policy-practice .oxfam.org.uk/publications/extreme-carbon-inequality-why-the-paris-climate -deal-must-put-the-poorest-lowes-582545.

138 **In the median household:** Jessica Semega et al., "Income and Poverty in the United States: 2019," United States Census Bureau (September 2020): 34, https://www.census.gov/content/dam/Census/library/publications/2020 /demo/p60-270.pdf.

138 **the income of two preschool teachers:** United States Bureau of Labor Statistics, "National Occupational Employment and Wage Estimates" (May 2019), https://www.bls.gov/oes/current/oes_nat.htm.

138 **an average of fifty (fifty!) tons:** Michael Eisenstein, "The Needs of the Many," *Nature* 551 (2017): figure 1, https://www.nature.com/articles/d41586 -017-07418-y.

138 **an astonishing ten thousand times more:** Stefan Gössling, "Celebri-ties, Air Travel, and Social Norms," *Annals of Tourism Research* 79 (2019), https://doi.org/10.1016/j.annals.2019.102775.

139 **budget to limit warming to 1.5°C:** Institute for Global Environmental Strategies (IGES), Aalto University, and D-mat ltd., *1.5-Degree Lifestyles: Tar-gets and Options for Reducing Lifestyle Carbon Footprints. Technical Report* (Hayama, Japan: Institute for Global Environmental Strategies, 2019).

139 **only 5 percent of European households:** Diana Ivanova and Richard Wood, "The Unequal Distribution of Household Carbon Footprints in Europe and Its Link to Sustainability," *Global Sustainability* 3, no. e18 (2020): 1–12, https://doi.org/10.1017/sus.2020.12.

139 **Sitra has profiled:** Jussi Impiö, Satu Lähteenoja, and Annina Orasmaa, "Pathways to 1.5-Degree Lifestyles by 2030," Sitra, 2020, https://www.sitra.fi/en/publications/pathways-to-1-5-degree-lifestyles-by-2030.

139 **60 percent less energy than today:** Joel Millward-Hopkins, "How 10 Billion People Could Live Well by 2050—Using As Much Energy As We Did 60 Years Ago," The Conversation, October 5, 2020, https://theconversation.com/how-10-billion-people-could-live-well-by-2050-using-as-much-energy-as-we-did-60-years-ago-146896.

139 **unnecessary for the global middle class:** IGES et al., *1.5-Degree Lifestyles,* v.

139 **reductions in luxury personal emissions:** Thomas Wiedmann et al., "Scientists' Warning on Affluence," *Nature Communications* 11, no. 3107 (2020), https://doi.org/10.1038/s41467-020-16941-y.

140 **cut *global* emissions by one-third:** Amy Goodman, "Climate Scientist: World's Richest Must Radically Change Lifestyles to Prevent Global Catastrophe," Democracy Now!, December 11, 2018, https://www.democracynow.org/2018/12/11/scientist_kevin_anderson_worlds_biggest_emitters.

140 **_emissions in places like the United States_:** Kevin Anderson, John F. Broderick, and Isak Stoddard, "A Factor of Two: How the Mitigation Plans of 'Climate Progressive' Nations Fall Far Short of Paris-Compliant Pathways," *Climate Policy* (2020), http://doi.org/10.1080/14693062.2020.1728209.

141 **human-warmed waters driving stronger hurricanes:** Robert McSweeney and Jocelyn Timperley, "Media Reaction: Hurricane Irma and Climate Change," Carbon Brief, December 9, 2017, https://www.carbonbrief.org/media-reaction-hurricane-irma-climate-change.

141 **Gaston Browne, the prime minister:** Zack Beauchamp, "What It's Like to Run a Country That Could Be Destroyed by Climate Change," Vox, February 13, 2018, https://www.vox.com/world/2018/2/13/17003008/climate-change-antigua-barbuda-gaston-browne.

142 **less happy and healthy:** Gibran Vita et al., "Happier with Less? Members of European Environmental Grassroots Initiatives Reconcile Lower Carbon Footprints with Higher Life Satisfaction and Income Increases," *Energy Research & Social Science* 60 (2020), https://doi.org/10.1016/j.erss.2019.101329; Rachel Sherman, *Uneasy Street: The Anxieties of Affluence* (Princeton, NJ: Princeton University Press, 2017).

143 **"The freedom to pollute":** Roy Scranton, "Raising a Daughter in a Doomed World," in *We're Doomed, Now What? Essays on War and Climate Change* (New York: Soho Press, 2018).

144 **personal and collective action are deeply entwined:** Shahzeen Z. Attari, David H. Krantz, and Elke U. Weber, "Climate Change Communicators' Carbon Footprints Affect Their Audience's Policy Support," *Climatic Change* 154, nos. 3–4 (2019): 529–45, https://doi.org/10.1007/s10584-019-02463-0; Leor Hackel and Gregg Sparkman, "Reducing Your Carbon Footprint Still Matters," *Slate,* October 26, 2018, https://slate.com/technology/2018/10/carbon-footprint-climate-change-personal-action-collective-action.html; Julia Steinberger, "An Audacious Toolkit: Actions Against Climate Breakdown.

Part 3: I Is for Individual," Medium, November 26, 2018, https://medium
.com/@JKSteinberger/an-audacious-toolkit-actions-against-climate-breakdown
-part-3-i-is-for-individual-f510ee035e13.

144   **businesses embrace personal footprints:** Alice Bell, "Beware Oil Execs in Environmentalists' Clothing—BP Could Derail Real Change," *The
Guardian,* February 14, 2020, https://www.theguardian.com/commentisfree
/2020/feb/14/oil-execs-environmentalists-bp-change-oil-climate.

144   **"The American way of life":** Marc Hudson, "George Bush Sr Could
Have Got in on the Ground Floor of Climate Action—History Would Have
Thanked Him," The Conversation, December 6, 2018, https://theconver
sation.com/george-bush-sr-could-have-got-in-on-the-ground-floor-of-climate
-action-history-would-have-thanked-him-108050.

144   **climate journalist Amy Harder:** Amy Harder, "Climate-Change Con-
fessions of an Energy Reporter," Axios, November 4, 2019, https://www.axios
.com/climate-change-confessions-energy-reporter-51b0fbf6-8c25-4616-88bd
-48aa16f149ae.html.

145   **will require carbon overconsumers:** B. Girod, D. P. Van Vuuren, and
E. G. Hertwich, "Global Climate Targets and Future Consumption Level:
An Evaluation of the Required GHG Intensity," *Environmental Research Letters* 8, no. 1 (2013): 014016; R. Carmichael, "Behaviour Change, Public Engagement and Net Zero. A Report for the Committee on Climate Change" (2019),
https://www.theccc.org.uk/publications/ and http://www.imperial.ac.uk
/icept/publications/.

146   **biggest bang for your climate buck:** I recommend Seth's excellent,
comprehensive but concise book quantifying personal emissions reductions:
Seth Wynes, *SOS: What You Can Do to Reduce Climate Change* (London:
Ebury Press, 2019).

146   **most people have no idea:** Seth Wynes, Jiaying Zhao, and Simon D.
Donner, "How Well Do People Understand the Climate Impact of Individual
Actions?" *Climatic Change* (2020), https://doi.org/10.1007/s10584-020
-02811-5.

146   **plant-based diet saves four times:** Seth Wynes and Kimberly Nicholas,
"The Climate Mitigation Gap: Education and Government Recommenda-
tions Miss the Most Effective Individual Actions," *Environmental Research
Letters* 12, no. 7 (2017), table 1, https://doi.org/10.1088/1748-9326/aa7541.

146   **more likely to spill over:** A. Gneezy et al., "Paying to Be Nice: Consis-
tency and Costly Prosocial Behavior," *Management Science* 58, no. 1 (2011):
179–87, https://doi.org/10.1287/mnsc.1110.1437.

146   **little evidence to suggest:** H. B. Truelove et al., "Positive and Negative
Spillover of Pro-environmental Behavior: An Integrative Review and Theo-
retical Framework," *Global Environmental Change* 29 (2014): 127–38; A.
Maki et al., "Meta-analysis of Pro-environmental Behaviour Spillover," *Nature
Sustainability* 2, no. 4 (2019): 307–15.

147   **may even backfire:** Maria Wolrath Söderberg and Nina Wormbs,
Grounded: Beyond Flygskam (Ixelles, Belgium: European Liberal Forum,
Fores, and the European Parliament, 2019).

147   **food goes to waste:** Project Drawdown, "Solution: Reduced Food Waste,"
https://www.drawdown.org/solutions/reduced-food-waste.

147   **twice as much carbon:** Wynes and Nicholas, "Climate Mitigation Gap,"
table 1.

148   **until it reaches a critical mass:** D. Centola, "The Spread of Behavior in an Online Social Network Experiment," *Science* 329, no. 5996 (2010): 1194–97.

148   **study headed by Jennifer Marlon:** Jennifer R. Marlon et al., "How Hope and Doubt Affect Climate Change Mobilization," *Frontiers in Communication* (2019), https://doi.org/10.3389/fcomm.2019.00020.

148   **80 percent in the United States and United Kingdom:** Solitaire Townsend, "Near 80% of People Would Personally Do as Much for Climate as They Have for Coronavirus," *Forbes,* June 1, 2020, https://www.forbes.com/sites/solitairetownsend/2020/06/01/near-80-of-people-would-ersonally-do-as-much-for-climate-as-they-have-for-coronavirus/.

148   **recent Swedish survey:** Niklas Birgetz, "Barn och unga om miljö," *Naturskyddsföreningen* (PowerPoint presentation, 2019), slides 13–14, https://www.naturskyddsforeningen.se/sites/default/files/dokument-media/novusundersokning_unga_om_miljo.pdf.

149   **internalizing knowledge of the seriousness:** Söderberg and Wormbs, *Grounded: Beyond Flygskam.*

149   **personal responsibility . . . seeing others actually act:** Marlon et al., "Hope and Doubt," 10–11. See also peer pressure/cultural change starting with individuals: Anna Kusmer, "Coronavirus Is Changing How People Think About Fighting Climate Change," Public Radio International, March 26, 2020, https://www.pri.org/stories/2020-03-26/coronavirus-changing-how-people-think-about-fighting-climate-change.

149   **avoiding flying:** Steve Westlake, "Climate Change: Yes, Your Individual Action Does Make a Difference," The Conversation, April 11, 2019, https://theconversation.com/climate-change-yes-your-individual-action-does-make-a-difference-115169.

149   **to misquote Gandhi:** Garson O'Toole, "Be the Change You Wish to See in the World," Quote Investigator, October 23, 2017, https://quoteinvestigator.com/2017/10/23/be-change/.

149   **inspire others to act:** Hackel and Sparkman, "Reducing Your Carbon Footprint."

149   **"personal rewards and these ripple effects":** Kim Cobb, "The Flying-Less Movement Webinar," October 30, 2019, quotes at 28:57 and 31:55, https://www.youtube.com/watch?v=yWas9U4Q_BM&t=1543s.

Chapter 9

151   **drowning in our things:** Jack Feuer, "The Clutter Culture," *UCLA Magazine,* July 1, 2012, http://magazine.ucla.edu/features/the-clutter-culture/print/.

151   **less visible consumption:** Christopher M. Jones and Daniel M. Kammen, "Quantifying Carbon Footprint Reduction Opportunities for U.S. Households and Communities," *Environmental Science and Technology* 45, no. 9 (2011): 4088–95, https://doi.org/10.1021/es102221h.

151   **unsustainable production and overconsumption:** Tobias D. Nielsen et al., "Politics and the Plastic Crisis: A Review Throughout the Plastic Life Cycle," *WIREs Energy and Environment* 9, no. 1 (2020): e36, https://doi.org/10.1002/wene.360.

152   **avoiding plastics saves way less emissions:** Zheng Jiajia and Sangwon Suh, "Strategies to Reduce the Global Carbon Footprint of Plastics," *Nature*

Climate Change 9, no. 5 (2019): 374–78, https://doi.org/10.1038/s41558-019 -0459-z.

152   **ten thousand avoided plastic water bottles:** Mike Berners-Lee, *How Bad Are Bananas? The Carbon Footprint of Everything* (London: Profile Books, 2010), 43 (carbon footprint of one plastic water bottle: 0.16 kg carbon dioxide equivalents, CO_2e); Seth Wynes and Kimberly Nicholas, "The Climate Mitigation Gap: Education and Government Recommendations Miss the Most Effective Individual Actions," *Environmental Research Letters* 12, no. 7 (2017), table 1, https://doi.org/10.1088/1748-9326/aa7541 (round-trip transatlantic flight: 1.6 tons CO_2e).

153   **to change their own behavior:** Shahzeen Z. Attari, David H. Krantz, and Elke U. Weber, "Statements About Climate Researchers' Carbon Footprints Affect Their Credibility and the Impact of Their Advice," *Climatic Change* 138, nos. 1–2 (2016): 325–38, https://doi.org/10.1007/s10584-016-1713-2.

153   **support ambitious climate policy:** Shahzeen Z. Attari, David H. Krantz, and Elke U. Weber, "Climate Change Communicators' Carbon Footprints Affect Their Audience's Policy Support," *Climatic Change* 154, nos. 3–4 (2019): 529–45, https://doi.org/10.1007/s10584-019-02463-0.

154   **"love miles":** George Monbiot, "On the Flight Path to Global Meltdown," *The Guardian*, September 21, 2006, https://www.theguardian.com/environ ment/2006/sep/21/travelsenvironmentalimpact.ethicalliving.

156   **only 11 percent . . . six or more flights:** Stefan Gössling and Andreas Humpe, "The Global Scale, Distribution and Growth of Aviation: Implications for Climate Change," *Global Environmental Change* 65 (2020): 102194, https://doi.org/10.1016/j.gloenvcha.2020.102194

156   **Our 2017 study:** Wynes and Nicholas, "Climate Mitigation Gap," figure 1.

157   **8 percent of global climate pollution:** Manfred Lenzen et al., "The Carbon Footprint of Global Tourism," *Nature Climate Change* 8 (2018): 522–28, https://doi.org/10.1038/s41558-018-0141-x.

157   **aviation alone can contribute 50 percent:** Ilona M. Otto et al., "Shift the Focus from the Super-Poor to the Super-Rich," *Nature Climate Change* 9 (2019): 82–84, figure 1, https://doi.org/10.1038/s41558-019-0402-3.

157   **two-thirds of his emissions:** Peter Kalmus, "How Far Can We Get Without Flying?" *Yes!*, February 11, 2016, https://www.yesmagazine.org/issues /life-after-oil/how-far-can-we-get-without-flying-20160211.

157   **warming equivalent to 7.2 percent:** D. S. Lee et al., "The Contribution of Global Aviation to Anthropogenic Climate Forcing for 2000 to 2018," *Atmospheric Environment* (2020), https://doi:10.1016/j.atmosenv.2020.117834.

157   **doubling in demand by 2037:** Jocelyn Timperley, "Corsia: The UN's Plan to 'Offset' Growth in Aviation Emissions After 2020," Carbon Brief, February 4, 2019, https://www.carbonbrief.org/corsia-un-plan-to-offset -growth-in-aviation-emissions-after-2020.

157   **disaster for the climate:** Kimberly Nicholas and Seth Wynes, "Flying Less Is Critical to a Safe Climate Future," *Public Administration Review*, Bully Pulpit Symposium on "The Green New Deal: Pathways to a Low Carbon Economy" (2019), https://www.publicadministrationreview.com/2019/07/16/gnd24/.

157   **no viable technical solutions:** Alice Bows-Larkin, "All Adrift. Aviation, Shipping, and Climate Change Policy," *Climate Policy* 15, no. 6 (2015): 681–702, https://doi.org/10.1080/14693062.2014.965125.

157 **22 percent of CO$_2$ emissions by 2050:** Martin Cames et al., "Emission Reduction Targets for International Aviation and Shipping," European Parliament, Directorate General for Internal Policies, https://www.europarl.europa .eu/thinktank/en/document.html?reference=IPOL_STU(2015)569964.

157 **especially by frequent flyers:** Nicholas and Wynes, "Flying Less Is Critical."

158 **only 58 percent of their trips:** Stefan Gössling et al., "Can We Fly Less? Evaluating the 'Necessity' of Air Travel," *Journal of Air Transport Management* 81 (2019), https://doi.org/10.1016/j.jairtraman.2019.101722.

158 **high-speed rail networks:** International Energy Agency, "The Future of Rail," January 2019, accessed July 25, 2020, https://www.iea.org/reports/the -future-of-rail.

158 **Aviation is currently heavily subsidized:** Stefan Gössling, Frank Fichert, and Peter Forsyth, "Subsidies in Aviation," *Sustainability* 9, no. 8 (2017): 1295, https://doi.org/10.3390/su9081295.

158 **European Citizens' Initiative:** ECI Fairosene, "Ending the Aviation Fuel Tax Exemption in Europe," European Citizens' Initiative, https://eci .ec.europa.eu/008/public/#/initiative.

158 **increased flight taxes:** Jonas Sonnenschein and Nora Smedby, "Designing Air Ticket Taxes for Climate Change Mitigation: Insights from a Swedish Valuation Study," *Climate Policy* 19, no. 5 (2018): 651–63, https://doi.org /10.1080/14693062.2018.1547678.

158 **frequent-flyer levy:** Leo Murray, "Public Attitudes to Tackling Aviation's Climate Change Impacts," 10:10 Climate Action, January 2019, https://s3-eu -west-1.amazonaws.com/media.afreeride.org/documents/Aviation_briefing_Jan 2019+FINAL.pdf.

158 **more than two-thirds:** Seth Wynes and Simon Donner, "Addressing Greenhouse Gas Emissions from Business-Related Air Travel at Public Institutions: A Case Study of the University of British Columbia," Pacific Institute for Climate Solutions, 2018, https://pics.uvic.ca/sites/default/files/AirTrav elWP_FINAL.pdf.

158 **leadership of the Tyndall Centre:** Corinne Le Quéré et al., "Towards a Culture of Low-Carbon Research for the 21st Century" (Tyndall Centre for Climate Change Research Working Paper 161, 2015).

158 **new travel policy:** Cecilia von Arnold, "LUCSUS Presents New Travel Policy to Reduce Work-Related Emissions," December 11, 2018, https://www .lucsus.lu.se/article/lucsus-presents-new-travel-policy-to-reduce-work-related -emissions.

160 **Transportation is the largest source:** US Environmental Protection Agency, "Sources of Greenhouse Gas Emissions," 2018, https://www.epa.gov /ghgemissions/sources-greenhouse-gas-emissions.

160 **gasoline for private cars:** Jones and Kammen, "Quantifying Carbon Footprint Reduction."

160 **transport still runs nearly 100 percent:** US Energy Information Administration, "Global Transportation Energy Consumption: Examination of Scenarios to 2040 Using ITEDD," US Department of Energy, 2017, figure 3, https://www.eia.gov/analysis/studies/transportation/scenarios/pdf/globaltrans portation.pdf.

160 **Long car trips . . . cause the most emissions:** Jillian Anable, "The Future of Transport After Covid 19: Everything Has Changed. Nothing Has Changed," UK Energy Research Centre, June 25, 2020, https://ukerc.ac.uk

/news/the-future-of-transport-after-covid-19-everything-has-changed-nothing
-has-changed/.

160 **especially for leisure:** Lena Smidfelt Rosqvist and Lena Winslott Hise-
lius, "Understanding High Car Use in Relation to Policy Measures Based on
Swedish Data," *Case Studies on Transport Policy* 7, no. 1 (2019): 28–36,
https://doi.org/10.1016/j.cstp.2018.11.004.

160 **big win for physical and mental health:** A. Martin, Y. Goryakin, and
M. Suhrcke, "Does Active Commuting Improve Psychological Wellbeing?
Longitudinal Evidence from Eighteen Waves of the British Household Panel
Survey," *Preventive Medicine* 69 (2014): 296–303, https://doi.org/10.1016
/j.ypmed.2014.08.023.

160 **feasible biking distance:** Federal Highway Administration, "2017 National Household Travel Survey, Popular Vehicle Trips Statistics," US Department of Transportation, Washington, DC, 2017, https://nhts.ornl.gov/vehicle
-trips.

160 **by putting people, not cars, at the center:** Giulio Mattioli et al., "The
Political Economy of Car Dependence: A Systems of Provision Approach,"
Energy Research & Social Science 66 (2020): 101486, https://doi.org/10.1016
/j.erss.2020.101486; Energy Innovation, "12 Green Guidelines," December 9,
2015, https://energyinnovation.org/publication/12-green-guidelines-2/; Stephen Higashide, *Better Buses, Better Cities: How to Plan, Run, and Win the
Fight for Effective Transit* (Washington, DC: Island Press, 2019).

161 **95 percent of the weight of a car:** Annie Lowrey, "Your Big Car Is
Killing Me," *Slate,* June 27, 2011, https://slate.com/business/2011/06/ameri
can-cars-are-getting-heavier-and-heavier-is-that-dangerous.html.

161 **emissions have more than canceled:** Laura Cozzi and Apostolos
Petropoulos, "Growing Preference for SUVs Challenges Emissions Reductions in Passenger Car Market," International Energy Agency, October 15,
2019, https://www.iea.org/commentaries/growing-preference-for-suvs-challenges
-emissions-reductions-in-passenger-car-market.

161 **If SUV drivers were a nation:** Niko Kommenda, "SUVs Second Biggest Cause of Emissions Rise, Figures Reveal," *The Guardian,* October 25,
2019, https://www.theguardian.com/environment/ng-interactive/2019/oct/25
/suvs-second-biggest-cause-of-emissions-rise-figures-reveal.

161 **about twice as much greenhouse gases:** J. Norman, H. L. MacLean,
and C. A. Kennedy, "Comparing High and Low Residential Density: Life-
Cycle Analysis of Energy Use and Greenhouse Gas Emissions," *Journal of
Urban Planning and Development* 132 (2006): 10–21.

161 **96 percent of their time parked:** Paul Barter, "'Cars Are Parked 95% of
the Time.' Let's Check!" *Reinventing Parking,* February 22, 2013, https://www
.reinventingparking.org/2013/02/cars-are-parked-95-of-time-lets-check.html.

161 **car users took three and a half times:** Felix Creutzig et al., "Fair
Street Space Allocation: Ethical Principles and Empirical Insights," *Transport
Reviews* (2020): 1–23, https://doi.org/10.1080/01441647.2020.1762795.

161 **sales of new electric cars:** Kyle Stock, "As Covid-19 Hits Electric
Vehicles, Some Thrive, Others Die," Bloomberg, May 19, 2020, https://www
.bloomberg.com/news/articles/2020-05-19/as-covid-19-hits-electric-vehicles-some
-thrive-others-die.

161 **electric vehicles today do reduce life-cycle emissions:** F. S. Kno-
bloch et al., "Net Emission Reductions from Electric Cars and Heat Pumps

in 59 World Regions over Time," *Nature Sustainability* 3, no. 6 (June 2020): 437–47, https://doi.org/10.1038/s41893-020-0488-7.

162   **half as much climate pollution:** Wynes and Nicholas, "Climate Mitigation Gap," figure 1.

162   **cannot achieve its necessary share:** Alexandre Milovanoff, I. Daniel Posen, and Heather L. MacLean, "Electrification of Light-Duty Vehicle Fleet Alone Will Not Meet Mitigation Targets," *Nature Climate Change* (2020), https://doi.org/10.1038/s41558-020-00921-7.

162   **UC Davis report:** Regina R. Clewlow and Gouri Shankar Mishra, "Disruptive Transportation: The Adoption, Utilization, and Impacts of Ride-Hailing in the United States," Institute of Transportation Studies, University of California, Davis, Research Report UCD-ITS-RR-17-07, October 2017, https://steps.ucdavis.edu/new-research-ride-hailing-impacts-travel-behavior/.

162   **Our 2018 study:** Seth Wynes et al., "Measuring What Works: Quantifying Greenhouse Gas Emission Reductions of Behavioural Interventions to Reduce Driving, Meat Consumption, and Household Energy Use," *Environmental Research Letters* 13, no. 11 (2018), https://doi.org/10.1088/1748-9326/aae5d7.

162   **experimented with going car-free:** Rachel Obordo, "'The Streets Are More Alive': Ghent Readers on a Car-Free City Centre," *The Guardian,* January 20, 2020, https://www.theguardian.com/environment/2020/jan/20/the-streets-are-more-alive-ghent-readers-on-a-car-free-city-centre.

163   **Reduced road vehicles contributed the biggest drop:** Corinne Le Quéré et al., "Temporary Reduction in Daily Global CO_2 Emissions During the COVID-19 Forced Confinement," *Nature Climate Change* 10, no. 7 (2020): figure 4, https://doi.org/10.1038/s41558-020-0797-x.

163   **Copenhagen was overrun by cars:** Jens Loft Rasmussen and Mai-Britt Kristensen, "Danish History: How Copenhagen Became Bike-Friendly Again," December 6, 2012, http://gridchicago.com/2012/danish-history-how-copenhagen-became-bike-friendly-again/.

163   **more bikes than cars:** Sean Fleming, "What Makes Copenhagen the World's Most Bike-Friendly City?" World Economic Forum, October 5, 2018, https://www.weforum.org/agenda/2018/10/what-makes-copenhagen-the-worlds-most-bike-friendly-city/.

163   **each kilometer cycled earns society $0.74:** Copenhagenize Design Co., "The Economic Benefits of Car-Free Streets," March 14, 2019, https://copenhagenize.eu/news-archive/2019/3/14/the-benefits-of-car-free-streets.

163–64   **25 percent of total greenhouse gases globally:** Intergovernmental Panel on Climate Change (IPCC), "Summary for Policymakers," in *Climate Change 2014: Mitigation of Climate Change. Contribution of Working Group III to the Fifth Assessment Report of the Intergovernmental Panel on Climate Change,* ed. O. Edenhofer et al. (Cambridge, UK: Cambridge University Press, 2014), figure SPM.2.

164   **electrify everything:** Steve Hanley, "Want to Limit Global Warming? Electrify Everything, Finds Study," CleanTechnica, April 16, 2019, https://cleantechnica.com/2019/04/16/want-to-limit-global-warming-electrify-everything-finds-study/.

164   **not 100 percent clean electricity, but 200 percent:** Leah Stokes, "Cleaning Up the Electricity System," *Democracy Journal,* no. 56 (2020), https://democracyjournal.org/magazine/56/cleaning-up-the-electricity-system/.

164   **seven months of driving:** Wynes and Nicholas, "Climate Mitigation Gap."

164   **rooftop solar panels:** Diana Ivanova et al., "Quantifying the Potential for Climate Change Mitigation of Consumption Options," *Environmental Research Letters* (2020): figure 5, https://doi.org/10.1088/1748-9326/ab8589.

164   **smaller-scale solutions:** C. A. Wilson et al., "Granular Technologies to Accelerate Decarbonization," *Science* 368, no. 6486 (2020): 36–39, https://doi.org/10.1126/science.aaz8060.

164   **most effective policies to date:** European Environment Agency, "More National Climate Policies Expected, but How Effective Are the Existing Ones?" November 27, 2019, figure 1, https://www.eea.europa.eu/themes/climate/national-policies-and-measures/more-national-climate-policies-expected.

165   **home energy use is the second-biggest:** Jones and Kammen, "Quantifying Carbon Footprint Reduction."

165   **save the most home energy:** Ivanova et al., "Quantifying the Potential for Climate Change."

165   **improving insulation and installing efficient windows:** É. Mata, A. Sasic Kalagasidis, and F. Johnsson, "Assessment of Retrofit Measures for Reduced Energy Use in Residential Building Stocks—Simplified Costs Calculation," Sustainable Building Conference 2010, figure 2, http://publications.lib.chalmers.se/records/fulltext/local_122318.pdf.

165   **electric heat pumps:** Ula Chrobak, "How Heat Pumps Can Help Fight Global Warming," *Popular Science,* March 3, 2020, https://www.popsci.com/story/environment/heat-pumps-emissions-climate-change/.

165   **biggest home energy user . . . turning off lights:** Shahzeen Z. Attari et al., "Public Perceptions of Energy Consumption and Savings," *Proceedings of the National Academy of Sciences of the United States of America* 107, no. 37 (2010): 16054–59.

166   **can leak a greenhouse gas:** Project Drawdown, "Refrigerant Management," https://drawdown.org/solutions/refrigerant-management.

166   **saves much less energy:** Wynes and Nicholas, "Climate Mitigation Gap," figure 1.

166   **exactly zero that examined reducing flying:** Wynes et al., "Measuring What Works."

166   **Malena Ernman publicly declared:** Malena Ernman, "Jorden behöver en överdos av godhet nu," *Expressen,* December 18, 2016, https://www.expressen.se/kultur/jorden-behover-en-overdos-av-godhet-nu/.

166–67   *flygskam* **(flight shame):** Institutet för språk och Folkminnen, "Flygskam," updated March 16, 2018, https://www.isof.se/sprak/nyord/nyord/aktuellt-nyord-2018/2018-03-16-flygskam.html.

167   **I'M TIRED OF SHOWING MY CHILD A DYING WORLD:** Jens Liljestrand, "Jag är trött på att visa mitt barn en döende värld," *Expressen Kultur,* January 13, 2018, https://www.expressen.se/kultur/jens-liljestrand/jag-ar-trott-pa-att-visa-mitt-barn-en-doende-varld/.

167   **lower price than research showed consumers were willing to pay:** Jonas Sonnenschein and Nora Smedby, "Designing Air Ticket Taxes for Climate Change Mitigation: Insights from a Swedish Valuation Study," *Climate Policy* 19, no. 5 (2018): 651–63, https://doi.org/10.1080/14693062.2018.1547678.

167   **"flight should pay its climate cost":** Naturskyddsföreningen, "Löften i elfte timmen: Granskning av partiernas miljöambitioner inför EU-valet 2019" [Eleventh-hour promises: Evaluation of the parties' environmental ambition before the 2019 EU election], May 2019, 28.

167   **Helsingborg added a 50 percent fee:** Marcus Friberg, "Klimatet viktigare än opinionssiffror," *Expressen,* May 24, 2018, https://www.expressen .se/kvallsposten/debatt-kvp/klimatet-viktigare-an-opinionssiffror/.

167   **international flights taken by Swedes:** Anneli Kamb and Jörgen Larsson, "Klimatpåverkan från Svenska befolkningens flygresor 1990–2017" [Climate impact of flights by the Swedish population, 1990–2017], Institutionen för Rymd-, geo- och miljövetenskap, Chalmers Tekniska Högskola, 2019, 4.

167   **Ryanair joined the top ten climate polluters:** "Ryanair One of Europe's Top Polluters, EU Data Suggests," BBC News, April 2, 2019, https:// www.bbc.com/news/business-47783992.

167   **carbon-offset schemes shown not to reduce emissions:** Jörgen Larsson et al., "International and National Climate Policies for Aviation: A Review," *Climate Policy* 19, no. 6 (2019): 787–99, https://doi.org/10.1080 /14693062.2018.1562871.

168   **ad encouraging customers to "fly responsibly":** Joanna Bailey, "KLM Urges Passengers to Fly Less in New Advert," Simple Flying, July 4, 2019, https://simpleflying.com/klm-fly-less/.

168   **devastated crop yields:** Petra Haupt, "Upprepad torka kan slå hårt mot lantbruket," Sveriges Radio, December 20, 2019, https://sverigesradio.se/sida /artikel.aspx?programid=96&artikel=7371988.

168   **forced to slaughter them early:** Helene Dauschy, "Sweden's Farmers Count Cost of Historic Drought," Phys.org, July 20, 2018, https://phys.org /news/2018-07-sweden-farmers-historic-drought.html.

168   **Countries as distant as Italy:** "In Pictures: Fighting the Swedish Wildfires," BBC, July 22, 2018, https://www.bbc.com/news/world-europe -44917545.

168   **"people care. . . . Most people would be prepared":** Maja Rosén, "Our Manifesto," We Stay on the Ground, n.d., https://westayontheground .blogspot.com/p/about.html.

168   **train from Stockholm to Venice:** "DN-tåget 2019," *Dagens Nyheter,* 2019, https://www.dn.se/om/dn-taget-2019/.

169   *flygskam* **debate to international coverage:** Jon Henley, "#stayonthe ground: Swedes Turn to Trains amid Climate 'Flight Shame,'" *The Guardian,* June 4, 2019, https://www.theguardian.com/world/2019/jun/04/stayonthe ground-swedes-turn-to-trains-amid-climate-flight-shame; "How Something Called Flygskam Is Hurting European Airlines," Bloomberg Businessweek, September 26, 2019, https://www.bloomberg.com/news/videos/2019-09-26 /how-something-called-flygskam-is-hurting-european-airlines-video; Sofia McFarland, "'Flight Shame' Comes to the U.S.—Via Greta Thunberg's Sailboat," *The Wall Street Journal,* August 30, 2019, https://www.wsj.com/arti cles/flight-shame-comes-to-the-u-s-via-sailboat-11567162801.

169   **desire to live in accordance with their values:** Maria Wolrath Söderberg and Nina Wormbs, *Grounded: Beyond Flygskam* (Ixelles, Belgium: European Liberal Forum, Fores, and the European Parliament, 2019).

169   **domestic train travel was up 11 percent:** "Resandet med SJ ökade med 11 procent 2019" [Travels with SJ up 11 percent in 2019], Travel News,

February 13, 2020, https://www.travelnews.se/sverige/sj/resandet-med-sj
-okade-med-11-procent-2019/.

169   **domestic flights were down 9 percent:** Transport Styrelsen [Swedish
Transport Authority], "Flygresandet minskade under 2019" [Air travel down
in 2019], January 22, 2020, https://www.transportstyrelsen.se/sv/Press/Press
meddelanden/2020/flygresandet-minskade-under-20192/.

169   **international flights also decreased slightly:** Swedavia Airports,
"Trafikstatistik på Swedavias flygplatser," https://www.swedavia.se/globalas
sets/statistik/swedavia_201912.pdf.

169   **14 percent of Swedes avoided flying:** "Klimat och miljö oroar nio av
tio svenskar," *Dagens Nyheter,* updated April 21, 2020, https://www.dn.se
/nyheter/sverige/klimat-och-miljo-oroar-nio-av-tio-svenskar/.

169   **only 25 percent of a population:** Damon Centola et al., "Experimental
Evidence for Tipping Points in Social Convention," *Science* 360, no. 6393
(2018): 1116–19, https://doi.org/10.1126/science.aas8827.

169   **all in the first year:** Maja Rosén, "International Campaigns," We Stay on
the Ground, n.d., https://westayontheground.blogspot.com/p/international
-campaigns.html; Maja Rosén, "GOTT NYTT FLYGFRITT ÅR!" Vi håller
oss på jorden, Facebook post, January 1, 2019, https://www.facebook.com
/vihallerosspajorden/posts/2236705943271021/.

170   **museum exhibit called Carbon Ruins:** "Carbon Ruins: The Fossil
Era," Climaginaries, https://www.climaginaries.org/exhibition.

170   **slated to succumb to the rising seas:** "Se hur vattnet sveper in över
Falsterbo" [See how the water sweeps over Falsterbo], *SVT Nyheter,* April 27,
2020, https://www.svt.se/nyheter/lokalt/skane/se-hur-vattnet-sveper-in
-over-falsterbonaset-om-havet-stiger.

Chapter 10

174   **In 1939, my father's parents:** Bill Lynch, "A Feather in Sonoma's Cap,"
Sonoma Index Tribune, November 23, 2016, https://www.sonomanews.com
/opinion/6346146-181/a-feather-in-sonomas-cap.

174   **optimized feed:** Turkey Farmers of Canada, "What Turkeys Eat," https://
www.turkeyfarmersofcanada.ca/on-the-farm/what-turkeys-eat-3/.

174   **as little as fourteen weeks:** Aviagen Turkeys, "Feeding Guidelines
for Nicholas and B.U.T. Heavy Lines," table 2A, http://www.aviagenturkeys
.com/uploads/2015/11/20/NU06%20Feeding%20Guidelines%20for%20Nicho
las%20&%20BUT%20Heavy%20Lines%20EN.pdf.

174   **Grandpa George's turkey:** "In 1957, Poultry Breeder George Nicholas
Marketed the First White Turkey," AP News, November 18, 1995, https://
apnews.com/2a94b6a354fdcbbc8fef9856cb268f28.

175   **analyzed more than twelve thousand scientific papers:** Lucia
Tamburino et al., "From Population to Production: 50 Years of Scientific Literature on How to Feed the World," *Global Food Security* 24 (2020): figure 3,
https://doi.org/10.1016/j.gfs.2019.100346.

175   **the problem is overproduction:** T. G. Benton and R. Bailey, "The Paradox of Productivity: Agricultural Productivity Promotes Food System Inefficiency," *Global Sustainability* 2, no. e6 (2019): 1–8, https://doi.org/10.1017
/sus.2019.3.

175   **Green Revolution tripled crop yields:** P. L. Pingali, "Green Revolution: Impacts, Limits, and the Path Ahead," *Proceedings of the National*

Academy of Sciences of the United States of America 109, no. 31 (2012): 12302–8, https://doi.org/10.1073/pnas.0912953109.

175 **two-thirds of calories:** Q. Ji, X. Xu, and K. Wang, "Genetic Transformation of Major Cereal Crops," *International Journal of Developmental Biology* 57, nos. 6–8 (2013): 495–508, https://doi.org/10.1387/ijdb.130244kw.

176 **Grandpa George's turkey is a prime example:** Mark Fritz, "Turkey's Story: A Fowl Tale of the Death of Diversity," *Los Angeles Times,* November 19, 1995, https://www.latimes.com/archives/la-xpm-1995-11-19-mn-4806-story.html.

176 **sixty-six countries currently lack:** Marianela Fader et al., "Spatial Decoupling of Agricultural Production and Consumption: Quantifying Dependences of Countries on Food Imports Due to Domestic Land and Water Constraints," *Environmental Research Letters* 8, no. 1 (2013), https://doi.org/10.1088/1748-9326/8/1/014046.

176 **comprehensive 2018 study in *Science*:** Joseph Poore and Thomas Nemecek, "Reducing Food's Environmental Impacts Through Producers and Consumers," *Science* 60, no. 6392 (2018): 987–92, https://doi.org/10.1126/science.aaq0216.

177 **analysis led by Emily Cassidy:** Emily S. Cassidy et al., "Redefining Agricultural Yields: From Tonnes to People Nourished per Hectare," *Environmental Research Letters* 8, no. 3 (2013), abstract and table 1, https://doi.org/10.1088/1748-9326/8/3/034015.

177 **half of today's cropland:** Peter Alexander et al., "Human Appropriation of Land for Food: The Role of Diet," *Global Environmental Change* 41 (2016): 88–98, https://doi.org/10.1016/j.gloenvcha.2016.09.005.

177 **most of the ice-free land:** Intergovernmental Panel on Climate Change (IPCC), "Summary for Policymakers," in *Climate Change and Land: An IPCC Special Report on Climate Change, Desertification, Land Degradation, Sustainable Land Management, Food Security, and Greenhouse Gas Fluxes in Terrestrial Ecosystems,* ed. P. R. Shukla et al. (2019), figure SPM.1, https://www.ipcc.ch/srccl/chapter/summary-for-policymakers/.

178 **extinctions *one thousand times* faster:** Mark D. A. Rounsevell et al., "A Biodiversity Target Based on Species Extinctions," *Science* 368, no. 6496 (2020): 1193–95, https://doi.org/10.1126/science.aba6592.

178 **up to 1 million species:** Intergovernmental Science-Policy Platform on Biodiversity and Ecosystem Services (IPBES), "Media Release: Nature's Dangerous Decline 'Unprecedented'; Species Extinction Rates 'Accelerating,'" May 2019, https://www.ipbes.net/news/Media-Release-Global-Assessment.

178 **80 percent of global deforestation:** Food and Agriculture Organization, *The Future of Food and Agriculture: Trends and Challenges* (Rome: Food and Agriculture Organization of the United Nations, 2017), http://www.fao.org/3/a-i6583e.pdf.

178 **eat *more* soy:** Richard Waite (@waiterich), "Nice visual explainer by @FCRNetwork on connections between soy, feed, and forests: https://foodsource.org.uk/building-blocks/soy-food-feed-and-land-use-change. It's counter-intuitive, but if you are an omnivore concerned about soy and deforestation you should eat *more* soy (e.g., tofu) and less meat from animals that are fed soy," February 4, 2020, 11:22 A.M., https://twitter.com/waiterich/status/1224775159250345984?s.

178   **The 2020 *Living Planet Report*:** World Wide Fund for Nature, *Living Planet Report 2020: Bending the Curve of Biodiversity Loss* (Gland, Switzerland: World Wide Fund for Nature, 2020), 16, 24.

179   **87 percent of water:** Kate A. Brauman et al., "Water Depletion: An Improved Metric for Incorporating Seasonal and Dry-Year Water Scarcity into Water Risk Assessments," *Elementa: Science of the Anthropocene* 4 (2016), figure 7 (raw data provided from the author), https://doi.org/10.12952/journal.elementa.000083.

179   **beef to bean burgers:** Helen Harwatt et al., "Substituting Beans for Beef as a Contribution Toward US Climate Change Targets," *Climatic Change* 143, nos. 1–2 (2017): 261–70, https://doi.org/10.1007/s10584-017-1969-1.

179   **bubble only thirty-five miles:** US Geological Survey, "All of Earth's Water in a Single Sphere!" July 16, 2019, https://www.usgs.gov/media/images/all-earths-water-a-single-sphere.

179   **Aral Sea:** This paragraph draws from: Philip Micklin, "The Aral Sea Disaster," *Annual Review of Earth and Planetary Sciences* 35, no. 1 (2007): 47–72, https://doi.org/10.1146/annurev.earth.35.031306.140120; NASA Earth Observatory, "World of Change: Shrinking Aral Sea," https://earthobservatory.nasa.gov/world-of-change/AralSea; Angela Fritz, "The Aral Sea Was Once the Fourth Largest Lake in the World. Watch It Dry Up," *The Washington Post*, September 30, 2014, https://www.washingtonpost.com/news/capital-weather-gang/wp/2014/09/30/the-aral-sea-was-once-the-fourth-largest-lake-in-the-world-watch-it-dry-up/.

179   **almost a quarter of all climate pollution:** IPCC, *Climate Change and Land*, "Summary for Policymakers."

180   **78 percent decrease in insect abundance:** S. Seibold et al., "Arthropod Decline in Grasslands and Forests Is Associated with Landscape-Level Drivers," *Nature* 574, no. 7780 (2019): 671–74, https://doi.org/10.1038/s41586-019-1684-3.

180   **53 percent decline in US grassland songbird populations:** Kenneth V. Rosenberg et al., "Decline of the North American Avifauna," *Science* 366 (2019): 120–24, https://doi.org/10.1126/science.aaw1313.

180   **one-third of food:** A. M. Klein et al., "Importance of Pollinators in Changing Landscapes for World Crops," *Proceedings of the Royal Society B: Biological Sciences* 274, no. 1608 (2007): 303–13, https://doi.org/10.1098/rspb.2006.3721.

180   **including a multitude of key crops:** Jeroen P. van der Sluijs and Nora S. Vaage, "Pollinators and Global Food Security: The Need for Holistic Global Stewardship," *Food Ethics* 1, no. 1 (2016): 75–91, https://doi.org/10.1007/s41055-016-0003-z.

180   **today most phosphorus:** James Elser and Elena Bennett, "A Broken Biogeochemical Cycle," *Nature* 478 (October 6, 2011): 29–31.

180   **threatening world food security:** D. Cordell, A. Turner, and J. Chong, "The Hidden Cost of Phosphate Fertilizers: Mapping Multi-stakeholder Supply Chain Risks and Impacts from Mine to Fork," *Global Change, Peace and Security* 27 (2015): 1–21.

181   **Steffen estimated that global phosphorus runoff:** Will Steffen et al., "Planetary Boundaries: Guiding Human Development on a Changing Planet," *Science* 347, no. 6223 (2015), figure 3, table 1, https://doi.org/10.1126/science1259855.

181 **largest cause of algal blooms:** Poore and Nemecek, "Reducing Food's Environmental Impacts."

181 **Most of the excessive fertilizer:** Steffen et al., "Planetary Boundaries," figure 2.

181 **three times as many people:** FAO, IFAD, UNICEF, WFP, and WHO, *The State of Food Security and Nutrition in the World 2020: Transforming Food Systems for Affordable Healthy Diets* (Rome: FAO, 2020), https://doi.org /10.4060/ca9692en.

181 **unhealthy diet was the largest mortality risk:** Ashkan Afshin et al., "Health Effects of Dietary Risks in 195 Countries, 1990–2017: A Systematic Analysis for the Global Burden of Disease Study 2017," *Lancet* 393, no. 10184 (2019): 1958–72, https://doi.org/10.1016/s0140-6736(19)30041-8.

181 **tobacco use was second:** Jeffrey D. Stanaway et al., "Global, Regional, and National Comparative Risk Assessment of 84 Behavioural, Environmental and Occupational, and Metabolic Risks or Clusters of Risks for 195 Countries and Territories, 1990–2017: A Systematic Analysis for the Global Burden of Disease Study 2017," *Lancet* 392, no. 10159 (2018): 1948, https://doi.org /10.1016/s0140-6736(18)32225-6.

181 **coronavirus pandemic had caused:** Max Roser, Hannah Ritchie, Esteban Ortiz-Ospina, and Joe Hasell, "Statistics and Research: Coronavirus Pandemic (COVID-19)," Our World in Data, December 30, 2020, https://our worldindata.org/coronavirus.

182 **plant-based diets shown to reduce:** D. Tilman and M. Clark, "Global Diets Link Environmental Sustainability and Human Health," *Nature* 515, no. 7528 (2014): 518–22, https://doi.org/10.1038/nature13959.

182 **largest source of antibiotic-resistant bacteria:** Melinda Wenner Moyer, "How Drug-Resistant Bacteria Travel from the Farm to Your Table," *Scientific American,* December 1, 2016, https://www.scientificamerican.com /article/how-drug-resistant-bacteria-travel-from-the-farm-to-your-table/.

182 **One Health:** Dylan Walsh, "Can One Health Save the World?" Ensia, June 24, 2013, https://ensia.com/features/can-one-health-save-the-world/.

182 **"lowest cost of liveweight production":** Aviagen Turkeys, "Feeding Guidelines," 4.

183 **Tobias Kuemmerle describes:** Tobias Kuemmerle, "Rewilding—A New Paradigm for Nature Conservation?," lecture, Biodiversity Theme Meeting, Biodiversity and Ecosystem Services in a Changing Climate, Lund University, February 28, 2020.

183 **reintroduction of wolves:** Brodie Farquhar, "Wolf Reintroduction Changes Ecosystem in Yellowstone," *Yellowstone National Park Trips,* June 30, 2020, https://www.yellowstonepark.com/things-to-do/wolf-reintroduction -changes-ecosystem.

183 **lost 94 percent of their capacity . . . "a world of giants":** Christopher E. Doughty et al., "Global Nutrient Transport in a World of Giants," *Proceedings of the National Academy of Sciences of the United States of America* 113, no. 4 (2016): 868–73, www.pnas.org/cgi/doi/10.1073/pnas.1502549112.

184 **pungent whale "poonado":** BBC Radio 5 Live, "Thar She Blows! Diver Caught in Whale 'Poonado,'" January 31, 2015, video, 2:01, https://www.you tube.com/watch?v=r4JJWuHmqPs.

184 **keep those deadly hunters inside:** Arie Trouwborst, Phillipa C. McCormack, and Elvira Martínez Camacho, "Domestic Cats and Their Impacts

on Biodiversity: A Blind Spot in the Application of Nature Conservation Law," *People and Nature* 2, no. 1 (2020): 235–50, https://doi.org/10.1002/pan3.10073.

184 **goofy-looking colorful clown collar:** S. K. Willson, I. A. Okunlola, and J. A. Novak, "Birds Be Safe: Can a Novel Cat Collar Reduce Avian Mortality by Domestic Cats (*Felis catus*)?" *Global Ecology and Conservation* 3 (2015): 359–66, https://doi.org/10.1016/j.gecco.2015.01.004.

184 **largest human-caused threat:** Scott R. Loss, Tom Will, and Peter P. Marra, "The Impact of Free-Ranging Domestic Cats on Wildlife of the United States," *Nature Communications* 4 (2013): 1396, https://doi.org/10.1038/ncomms2380.

185 **just about enough phosphorus:** Usman Akram et al., "Optimizing Nutrient Recycling from Excreta in Sweden and Pakistan: Higher Spatial Resolution Makes Transportation More Attractive," *Frontiers in Sustainable Food Systems* 3 (2019), https://doi.org/10.3389/fsufs.2019.00050.

185 **Sydney, Australia, she finds:** Geneviève S. Metson et al., "Mapping Phosphorus Hotspots in Sydney's Organic Wastes: A Spatially Explicit Inventory to Facilitate Urban Phosphorus Recycling," *Journal of Urban Ecology* 4, no. 1 (2018), https://doi.org/10.1093/jue/juy009.

185 **rethink the purpose:** Walter Willett et al., "Food in the Anthropocene: The EAT–*Lancet* Commission on Healthy Diets from Sustainable Food Systems," *Lancet* 393, no. 10170 (2019): 447–92, https://doi.org/10.1016/S0140-6736(18)31788-4.

185 **improved land and agriculture techniques:** Project Drawdown, "Food, Agriculture, and Land Use," https://drawdown.org/sectors/food-agriculture-land-use; "Regenerative Annual Cropping," https://drawdown.org/solutions/regenerative-annual-cropping.

185 **$1 million *every minute*:** Damian Carrington, "$1M a Minute: The Farming Subsidies Destroying the World—Report," *The Guardian,* September 16, 2019, https://www.theguardian.com/environment/2019/sep/16/1m-a-minute-the-farming-subsidies-destroying-the-world.

185 **more than 80 percent of farm subsidies:** European Commission, "Direct Payments to Agricultural Producers—Graphs and Figures: Financial Year 2017," 2017, https://ec.europa.eu/agriculture/sites/agriculture/files/cap-funding/beneficiaries/direct-aid/pdf/direct-aid-report-2017_en.pdf.

186 **EU farm subsidies were misspent:** Murray W. Scown, Mark V. Brady, and Kimberly Nicholas, "Billions in Misspent EU Agricultural Subsidies Could Support the Sustainable Development Goals," *One Earth* 3, no. 2 (2020): 237–50, https://doi.org/10.1016/j.oneear.2020.07.011.

186 **start turning around the screaming plummet of biodiversity:** D. Leclère et al., "Bending the Curve of Terrestrial Biodiversity Needs an Integrated Strategy," *Nature* (2020), https://doi.org/10.1038/s41586-020-2705-y.

186 **food system compatible with a living planet:** M. Springmann et al., "Options for Keeping the Food System Within Environmental Limits," *Nature* 562, no. 7728 (2018): 519–25, https://doi.org/10.1038/s41586-018-0594-0.

186 **plant-based diets were the highest-impact:** Seth Wynes and Kimberly Nicholas, "The Climate Mitigation Gap: Education and Government Recommendations Miss the Most Effective Individual Actions," *Environmental Research Letters* 12, no. 7 (2017), table 1, https://doi.org/10.1088/1748-9326/aa7541.

186   **menu of twenty-two solutions was needed:** Janet Ranganathan et al., "How to Sustainably Feed 10 Billion People by 2050, in 21 Charts," World Resources Institute, December 5, 2018, https://www.wri.org/blog/2018/12 /how-sustainably-feed-10-billion-people-2050-21-charts.

186   **biggest environmental benefit is cutting beef:** Daisy Dunne, "Interactive: What Is the Climate Impact of Eating Meat and Dairy?" Carbon Brief, September 14, 2020, https://interactive.carbonbrief.org/what-is-the-climate -impact-of-eating-meat-and-dairy/.

186   **the 2°C target without reducing today's meat:** Fredrik Hedenus, Stefan Wirsenius, and Daniel J. A. Johansson, "The Importance of Reduced Meat and Dairy Consumption for Meeting Stringent Climate Change Targets," *Climatic Change* 124 (2014): 79–91, https://doi.org/10.1007/s10584-014 -1104-5.

187   **even the best-raised animals:** Poore and Nemecek, "Reducing Food's Environmental Impacts," figure 1.

187   **transport is usually a small fraction:** Sonja J. Vermeulen, Bruce M. Campbell, and John S. I. Ingram, "Climate Change and Food Systems," *Annual Review of Environment and Resources* 37, no. 1 (2012), table 1, https:// doi.org/doi:10.1146/annurev-environ-020411-130608.

187   **requires 80 percent less livestock:** J. O. Karlsson et al., "Designing a Future Food Vision for the Nordics Through a Participatory Modeling Approach," *Agronomy for Sustainable Development* 38, no. 6 (2018): 59, https:// doi.org/10.1007/s13593-018-0528-0.

187   **"express its creaturely character":** Michael Pollan, "An Animal's Place," *The New York Times Magazine,* November 10, 2002, https://michael pollan.com/articles-archive/an-animals-place/.

187   **twice as much protein:** Sophie Egan, "How Much Protein Do We Need?" *The New York Times,* July 28, 2017, https://www.nytimes.com/2017/07/28 /well/eat/how-much-protein-do-we-need.html.

187   **EAT-*Lancet* authors found:** Willett, "Food in the Anthropocene," figure 1.

188   **cow milk:** Clara Guibougt and Helen Briggs, "Climate Change: Which Vegan Milk Is Best?" BBC News, February 29, 2019, https://www.bbc.com /news/science-environment-46654042.

188   **Planetary Health Plate:** Harvard T. H. Chan School of Public Health, "Plate and the Planet," https://www.hsph.harvard.edu/nutritionsource/sus tainability/plate-and-planet/.

188   ***Grist's* Eve Andrews suggests:** Eve Andrews, "Can I Have My Climate-Friendly Seafood and Eat It, Too?" *Grist,* July 11, 2019, https://grist.org/food /can-i-have-my-climate-friendly-seafood-and-eat-it-too/.

188   **food impacts on biodiversity:** Emma Moberg et al., "Benchmarking the Swedish Diet Relative to Global and National Environmental Targets— Identification of Indicator Limitations and Data Gaps," *Sustainability* 12, no. 4 (2020), supplementary table 2, https://doi.org/10.3390/su12041407.

189   **first on the menu and in buffets:** Seth Wynes et al., "Measuring What Works: Quantifying Greenhouse Gas Emission Reductions of Behavioural Interventions to Reduce Driving, Meat Consumption, and Household Energy Use," *Environmental Research Letters* 13, no. 11 (2018), https://doi.org/10 .1088/1748-9326/aae5d7.

189   **emphasizing their delicious taste:** B. P. Turnwald et al., "Increasing Vegetable Intake by Emphasizing Tasty and Enjoyable Attributes: A

Randomized Controlled Multisite Intervention for Taste-Focused Labeling," *Psychological Science* 30, no. 11 (2019): 1603–15, http://doi.org/10.1177 /0956797619872191.

189–90 **meeting the vegetable consumption of city dwellers:** F. Martel-lozzo et al., "Urban Agriculture: A Global Analysis of the Space Constraint to Meet Urban Vegetable Demand," *Environmental Research Letters* 9, no. 6 (2014): 064025, https://doi.org/10.1088/1748-9326/9/6/064025.

190 **planting food trees in cities:** Kyle H. Clark and Kimberly Nicholas, "Introducing Urban Food Forestry: A Multifunctional Approach to Increase Food Security and Provide Ecosystem Services," *Landscape Ecology* 28, no. 9 (2013): 1649–69, https://doi.org/10.1007/s10980-013-9903-z.

Chapter 11

191 **Sonoma was enveloped in smoke:** Erin Brodwin, "The Devastating California Wine Country Fires Have Made the Air More Toxic than Bei-jing's—and It's Showing No Signs of Stopping," *Business Insider*, October 13, 2017, https://www.businessinsider.com/california-wildfires-dirtiest-worst -air-health-2017-10.

192 **artifacts from the historical landmarks:** Marissa Lang, "California Removes Artifacts from Historic Spanish Mission as Fire Nears," *SFGate*, October 11, 2017, https://www.sfgate.com/news/article/California-removes -artifacts-from-historic-12271754.php#photo-14332574.

193 **Dozens of people died:** Alex Emslie, "October Fires' 44th Victim: A Creative, Globetrotting Engineer with 'the Kindest Heart,'" KQED, Novem-ber 28, 2017, https://www.kqed.org/news/11633757/october-fires-44th-victim -a-creative-globetrotting-engineer-with-the-kindest-heart.

193 **including an elderly couple:** Kimberly Veklerov and Marissa Lang, "Married 75 Years, Couple Dies Together in Napa Fire," *SFGate*, October 10, 2017, https://www.sfgate.com/bayarea/article/Elderly-couple-identified-as -first-casualties-of-12266922.php.

193 **The combination of hotter spring:** Michael Goss et al., "Climate Change Is Increasing the Likelihood of Extreme Autumn Wildfire Condi-tions Across California," *Environmental Research Letters* 15, no. 9 (2020), https://doi.org/10.1088/1748-9326/ab83a7.

193 **acted as carbon sources rather than sinks:** Patrick Gonzalez et al., "Aboveground Live Carbon Stock Changes of California Wildland Ecosystems, 2001–2010," *Forest Ecology and Management* 348 (2015): 68–77, https://doi.org /10.1016/j.foreco.2015.03.040; B. M. Sleeter et al., "Effects of 21st-Century Climate, Land Use, and Disturbances on Ecosystem Carbon Balance in Cali-fornia," *Global Change Biology* 25, no. 10 (2019): 3334–53, https://doi.org /10.1111/gcb.14677.

193 **nearly doubled the area of forest fires:** T. John Abatzoglou and A. Park Williams, "Impact of Anthropogenic Climate Change on Wildfire Across Western US Forests," *Proceedings of the National Academy of Sciences of the United States of America* 18, no. 42 (2016): 11770–75, https://doi.org/10.1073 /pnas.1607171113.

193 **doubled the days of extreme autumn fire:** Goss et al., "Climate Change Is Increasing."

195 **often highly subsidized:** Stefan Gössling, Frank Fichert, and Peter For-syth, "Subsidies in Aviation," *Sustainability* 9, no. 8 (2017), https://doi.org

/10.3390/su9081295; Marine Formentini, *Growing Better: Ten Critical Transitions to Transform Food and Land Use* (2019), https://www.foodandlandusecoalition.org/wp-content/uploads/2019/09/FOLU-GrowingBetter-GlobalReport.pdf.

195 **recall that the purpose of the economy:** Kate Raworth, *Doughnut Economics: Seven Ways to Think Like a 21st-Century Economist* (London: Penguin Random House, 2018).

195 **study led by Julia Steinberger:** Julia K. Steinberger, William F. Lamb, and Marco Sakai, "Your Money or Your Life? The Carbon-Development Paradox," *Environmental Research Letters* 15, no. 4 (2020), https://doi.org/10.1088/1748-9326/ab7461.

195 **"optimal" level of warming is 3.5°C:** Jason Hickel, "The Nobel Prize for Climate Catastrophe," *Foreign Policy,* December 6, 2018, https://foreignpolicy.com/2018/12/06/the-nobel-prize-for-climate-catastrophe/.

196 **study by Frances Moore and Delavane Diaz:** Frances C. Moore and Delavane B. Diaz, "Temperature Impacts on Economic Growth Warrant Stringent Mitigation Policy," *Nature Climate Change* 5, no. 2 (2015): 127–31, https://doi.org/10.1038/nclimate2481.

196 **23 percent loss of global GDP:** M. Burke, S. M. Hsiang, and E. Miguel, "Global Non-Linear Effect of Temperature on Economic Production," *Nature* 527, no. 7577 (2015): 235–39, https://doi.org/10.1038/nature15725.

196 **pandemic would lead to a 4.4 percent loss:** International Monetary Fund, "World Economic Outlook: A Long and Difficult Ascent," October 2020: table 1.1, https://www.imf.org/en/Publications/WEO/Issues/2020/09/30/world-economic-outlook-october-2020.

196 **"economies that make us thrive":** Raworth, *Doughnut Economics*, 30.

196 **Such an economic system:** H. E. Daly, "Allocation, Distribution, and Scale: Towards an Economics That Is Efficient, Just, and Sustainable," *Ecological Economics* 6 (1992): 185–93.

197 **would "asphyxiate" the economy:** "What They Don't Tell You About Climate Change," *The Economist,* November 16, 2017, https://www.economist.com/news/leaders/21731397-stopping-flow-carbon-dioxide-atmosphere-not-enough-it-has-be-sucked-out/.

198 **some losses are irreplaceable:** Roz Pidcock and Sophie Yeo, "Explainer: Dealing with the 'Loss and Damage' Caused by Climate Change," Carbon Brief, May 9, 2017, https://www.carbonbrief.org/explainer-dealing-with-the-loss-and-damage-caused-by-climate-change.

198 **fossil phaseout could save 3.6 million lives:** J. Lelieveld et al., "Effects of Fossil Fuel and Total Anthropogenic Emission Removal on Public Health and Climate," *Proceedings of the National Academy of Sciences of the United States of America* 116, no. 15 (2019): 7192–97, https://doi.org/10.1073/pnas.1819989116.

198 **almost $500 billion:** Marlowe Hood, "2019 Fossil Fuel Subsidies Nearly $500 bn: OECD/IEA," AFP, June 9, 2020, https://news.yahoo.com/2019-fossil-fuel-subsidies-nearly-500-bn-oecd-110125899.html.

198–99 **6.5 percent of global GDP:** David Coady et al., "Global Fossil Fuel Subsidies Remain Large: An Update Based on Country-Level Estimates" (International Monetary Fund Working Paper, May 2019), https://www.imf.org/en/Publications/WP/Issues/2019/05/02/Global-Fossil-Fuel-Subsidies-Remain-Large-An-Update-Based-on-Country-Level-Estimates-46509.

199 **coalition working to end fossil subsidies:** MISSION2020, "Inspirational Times: Tracking Progress Toward the 2020 Climate Turning Point," https://mission2020.global/wp-content/uploads/Mission2020_DavosEdition_web.pdf.

199 **Even in Sweden:** Fossilfritt Sverige, "Fossil Free Sweden," http://fossilfritt-sverige.se/in-english/; "Energy Use in Sweden," https://sweden.se/nature/energy-use-in-sweden/.

199 **three times more public money:** "Staten Subventionerar Klimatfarliga Utsläpp" ["The state subsidizes climate-harming emissions"], *Svenska Dagbladet*, March 5, 2018, https://www.svd.se/staten-subventionerar-klimat-farliga-utslapp.

199 **Direct global fossil fuel subsidies:** Michael Taylor, "Energy Subsidies: Evolution in the Global Energy Transformation to 2050" (staff technical paper, International Renewable Energy Agency, Abu Dhabi, 2020), https://www.irena.org/-/media/Files/IRENA/Agency/Publication/2020/Apr/IRENA_Energy_subsidies_2020.pdf.

199 **A 2020 analysis:** International Renewable Energy Agency, *Renewable Power Generation Costs in 2019* (Abu Dhabi: International Renewable Energy Agency, 2020): 11, 13, 15, 22–23, 37–38, https://www.irena.org/-/media/Files/IRENA/Agency/Publication/2020/Jun/IRENA_Power_Generation_Costs_2019.pdf.

200 **2020 *Nature* article:** Niklas Höhne et al., "Emissions: World Has Four Times the Work or One-Third of the Time," *Nature* 579 (2020): 25–28, https://www.nature.com/articles/d41586-020-00571-x.

200 **economy-wide carbon price:** High-Level Commission on Carbon Prices, *Report of the High-Level Commission on Carbon Prices* (Washington, DC: World Bank, 2017), https://www.carbonpricingleadership.org/report-of-the-highlevel-commission-on-carbon-prices.

200 **one-fifth of global carbon emissions:** World Bank, *State and Trends of Carbon Pricing 2020* (Washington, DC: World Bank, 2020), https://openknowledge.worldbank.org/handle/10986/33809.

200 **average price . . . of just $8/ton:** The Organization for Economic Cooperation and Development (OECD) reports a gap of 76.5 percent between the actual carbon prices and what they estimate as the real climate cost of €30/ton: 23.5 percent of €30 equals €7, or about US$8: OECD, "Few Countries Are Pricing Carbon High Enough to Meet Climate Targets," September 18, 2019, http://www.oecd.org/environment/few-countries-are-pricing-carbon-high-enough-to-meet-climate-targets.htm.

200 **Only 10 percent of emissions are priced:** "Fiscal Policy and Carbon Pricing," slide 17, in "UNEP Emissions Gap Report 2018" (presentation at Intergovernmental Panel on Climate Change [IPCC] side event at the 24th Conference of the Parties to the United Nations Framework Convention on Climate Change (COP24), Katowice, Poland, December 5, 2018), https://www.ipcc.ch/site/assets/uploads/2018/12/UNEP-1.pdf.

200 **at best a marginal dent:** E. Tvinnereim and M. Mehling, "Carbon Pricing and Deep Decarbonization," *Energy Policy* 121 (2018): 185–89, https://doi.org/10.1016/j.enpol.2018.06.020.

201 **too many permits to pollute:** "ETS, RIP?" *The Economist*, April 20, 2013, https://www.economist.com/finance-and-economics/2013/04/20/ets-rip.

201 **far too cheap to keep polluting:** OECD, "Effective Carbon Rates 2018: Pricing Carbon Emissions Through Taxes and Emissions Trading," 2018, http://www.oecd.org/tax/effective-carbon-rates-2018-9789264305304-en.htm.

201　**the world's highest carbon price:** World Bank, *State and Trends of Carbon Pricing 2020.*

201　**declined about 1 percent per year:** World Bank, "When It Comes to Emissions, Sweden Has Its Cake and Eats It Too," May 16, 2016, https://www.worldbank.org/en/news/feature/2016/05/16/when-it-comes-to-emissions-sweden-has-its-cake-and-eats-it-too.

201　**needs to reduce emissions . . . 7.6 percent:** United Nations Environment Programme (UNEP), *Emissions Gap Report 2019* (Nairobi: UNEP, 2019).

201　**increasingly advocating for a carbon price:** George P. Shultz and Ted Halstead, "The Winning Conservative Climate Solution," *The Washington Post,* January 16, 2020, https://www.washingtonpost.com/opinions/the-winning-republican-climate-solution-carbon-pricing/2020/01/16/d6921dc0-38 7b-11ea-bf30-ad313e4ec754_story.html.

201　**Oil companies are shifting strategy:** Umair Irfan, "Exxon Is Lobbying for a Carbon Tax. There Is, Obviously, a Catch," Vox, October 18, 2018, https://www.vox.com/2018/10/18/17983866/climate-change-exxon-carbon-tax-lawsuit.

202　**not a panacea:** Tvinnereim and Mehling, "Carbon Pricing and Deep Decarbonization"; D. Burtraw, A. Keyes, and L. Zetterberg, "Companion Policies Under Capped Systems and Implications for Efficiency—The North American Experience and Lessons in the EU Context," Resources for the Future, 2018, https://media.rff.org/documents/RFF-Rpt-Companion20Policies20and 20Carbon20Pricing_0.pdf.

202　**no silver bullets:** Bill McKibben, "Welcome to the Climate Crisis: How to Tell Whether a Candidate Is Serious About Combating Global Warming," *The Washington Post,* May 27, 2006, https://www.washingtonpost.com /archive/opinions/2006/05/27/welcome-to-the-climate-crisis-span-classbank headhow-to-tell-whether-a-candidate-is-serious-about-combating-global-warm ingspan/26b2ac5a-a4a3-46ff-b214-3fc07a3a5ab3/.

202　**nowhere near enough on its own:** Kevin M. Kennedy, "Putting a Price on Carbon: Evaluating a Carbon Price and Complementary Policies for a 1.5°C World," World Resources Institute Issue Brief, September 2019, https://files.wri.org/s3fs-public/putting-price-carbon.pdf.

202　**economists argue that redistributing revenue:** Franziska Funke and Linus Mattauch, "Why Is Carbon Pricing in Some Countries More Successful Than in Others?" Our World In Data, August 10, 2018, https://ourworldin data.org/carbon-pricing-popular.

202　**regulation to require energy:** European Environment Agency, "More National Climate Policies Expected, but How Effective Are the Existing Ones?" November 27, 2019, figure 3, https://www.eea.europa.eu/themes/climate /national-policies-and-measures/more-national-climate-policies-expected; J. Rogelj et al., "Mitigation Pathways Compatible with 1.5°C in the Context of Sustainable Development," in *Global Warming of 1.5°C: An IPCC Special Report on the Impacts of Global Warming of 1.5°C Above Pre-Industrial Levels and Related Global Greenhouse Gas Emission Pathways, in the Context of Strengthening the Global Response to the Threat of Climate Change, Sustainable Development, and Efforts to Eradicate Poverty,* ed. V. Masson-Delmotte et al. (Geneva, Switzerland: IPCC, 2018), 148.

203　**restrict and finally stop:** Peter Erickson, Michael Lazarus, and Georgia Piggot, "Limiting Fossil Fuel Production as the Next Big Step in Climate

Policy," *Nature Climate Change* 8, no. 12 (2018): 1037–43, https://doi.org
/10.1038/s41558-018-0337-0.

203   **fossil fuel bans:** Fergus Green, "The Logic of Fossil Fuel Bans," *Nature
       Climate Change* 8, no. 6 (2018): 449–51, https://doi.org/10.1038/s41558
       -018-0172-3.

203   **recommends the Swedish government set a stop date:** Swedish
       Climate Policy Council, "2019 Report of the Swedish Climate Policy Council,"
       Report 2, 2019, https://www.klimatpolitiskaradet.se/wp-content/uploads
       /2019/09/climatepolicycouncilreport2.pdf.

204   **starting to refuse to fund fossil enterprises:** Esteban Duarte, "Al-
       berta Rejects Oil Sands Stigma After Sweden Dumps Bonds," BNN Bloom-
       berg, November 15, 2019, https://www.bnnbloomberg.ca/alberta-rejects-oil
       -sands-stigma-after-sweden-dumps-bonds-1.1348677; Julia Kollewe, "Coal
       Power Becoming 'Uninsurable' as Firms Refuse Cover," *The Guardian*, De-
       cember 2, 2019, https://www.theguardian.com/environment/2019/dec/02
       /coal-power-becoming-uninsurable-as-firms-refuse-cover; Tim Gray, "Funds
       That Can Put Your Investments on a Low-Carbon Diet," *The New York Times*,
       October 13, 2017, https://www.nytimes.com/2017/10/13/business/mutfund
       /mutual-funds-low-carbon.html.

204   **As Yale scientist Jennifer Marlon writes:** Jennifer Marlon,
       "7 Ways You're Already Paying for Climate Change," *Barron's*, September 13,
       2020, https://www.barrons.com/articles/7-ways-youre-already-paying-for
       -climate-change-51599995430.

204   **study led by David McCollum:** David L. McCollum et al., "Energy In-
       vestment Needs for Fulfilling the Paris Agreement and Achieving the Sus-
       tainable Development Goals," *Nature Energy* 3, no. 7 (2018): 589–99, https://
       doi.org/10.1038/s41560-018-0179-z.

204   **According to the IPCC, investing 2.5 percent:** IPCC, "Summary
       for Policymakers," in *Global Warming of 1.5°C*, section D5.3, https://www
       .ipcc.ch/sr15/chapter/spm/.

205   **last decade from economies of scale:** Goksin Kavlak, James McNer-
       ney, and Jessika E. Trancik, "Evaluating the Causes of Cost Reduction in
       Photovoltaic Modules," *Energy Policy* 123 (2018): 700–10, https://doi.org
       /10.1016/j.enpol.2018.08.015.

205   **Germany's renewable electricity policy:** J. Lipp, "Lessons for Effec-
       tive Renewable Electricity Policy from Denmark, Germany and the United
       Kingdom," *Energy Policy* 35 (2007): 5481–95, https://doi.org/10.1016/j.enpol
       .2007.05.015.

205   **decentralized, local community-based energy:** David Roberts,
       "Wildfires and Blackouts Mean Californians Need Solar Panels and Mi-
       crogrids," Vox, October 28, 2019, https://www.vox.com/energy-and-environ
       ment/2019/10/28/20926446/california-grid-distributed-energy.

205   **smaller-scale, lower-cost "tiny tech":** C. A. Wilson et al., "Granular
       Technologies to Accelerate Decarbonization," *Science* 368, no. 6486 (2020):
       36–39, https://doi.org/10.1126/science.aaz8060; Walter Beckwith, "Tiny Tech
       Needed for Rapid Progress Towards Emissions Targets," American Associa-
       tion for the Advancement of Science, April 2, 2020, https://www.aaas
       .org/news/tiny-tech-needed-rapid-progress-towards-emissions-targets.

206   **assessment of how the mega-banks:** Rainforest Action Network,
       "Banking on Climate Change: Fossil Fuel Finance Report Card 2019," 2019,

8, https://www.ran.org/wp-content/uploads/2019/03/Banking_on_Climate _Change_2019_vFINAL1.pdf.

206 **funding the Dakota Access Pipeline:** Emily Fuller, "How to Contact the 3 Dozen Banks Still Backing Dakota Access Pipeline Companies," *Yes!*, September 29, 2016, http://www.yesmagazine.org/people-power/how -to-contact-the-17-banks-funding-the-dakota-access-pipeline-20160929.

206 **among the top funders:** Rainforest Action Network, "Banking on Climate Change," 14.

207 **climate-destroying mega-bank, UBS:** Rainforest Action Network, "Banking on Climate Change," 14.

207 **along with their investments in dirty energy:** Rainforest Action Network, "Banking on Climate Change."

208 **"take away the social license":** Bill McKibben (@billmckibben), "When people set out to . . . build the divestment movement a decade ago, one goal was to 'take away the social license' of the fossil fuel industry. It's built to the point where in the last 2 weeks the Pope and the Queen have joined in; thanks to all who made it all happen!" Twitter, June 26, 2020, 5:16 A.M., https://twitter.com/billmckibben/status/1276489610554720256?s.

208 *New Yorker* **article:** Carolyn Kormann, "The Divestment Movement to Combat Climate Change Is All Grown Up," *The New Yorker,* September 14, 2018, https://www.newyorker.com/news/dispatch/the-divestment-move ment-to-combat-climate-change-is-all-grown-up.

208 **biggest fund manager, BlackRock:** Joanna Partridge, "World's Biggest Fund Manager Vows to Divest from Thermal Coal," *The Guardian,* January 14, 2020, https://www.theguardian.com/business/2020/jan/14/blackrock -says-climate-crisis-will-now-guide-its-investments.

208 **I've long supported the campaign:** Kimberly Nicholas, "Lund University Faculty: Time to #Divest from Fossil Fuels" (speech, February 14, 2015), http://www.kimnicholas.com/blog/lund-university-faculty-time-to-divest-from -fossil-fuels.

208 **don't truly benefit local communities:** Wim Carton, "Rendering Local: The Politics of Differential Knowledge in Carbon Offset Governance," *Annals of the American Association of Geographers* (2020): 1–16, https://doi .org/10.1080/24694452.2019.1707642.

209 **only 2 percent . . . compared with 85 percent:** Martin Cames et al., *How Additional Is the Clean Development Mechanism?* (Berlin: Stockholm Environment Institute and Öko-Institut, 2016), https://ec.europa.eu/clima /sites/clima/files/ets/docs/clean_dev_mechanism_en.pdf.

209 **a "future box":** Leave It in the Ground (LINGO), "Future Box," http://leave -it-in-the-ground.org/future-box.

210 **manage the changes we cannot avoid:** I believe I first heard this idea from Jane Lubchenco.

210 **2014 study, wildfire scientists:** D. E. Calkin et al., "How Risk Management Can Prevent Future Wildfire Disasters in the Wildland-Urban Interface," *Proceedings of the National Academy of Sciences of the United States of America* 111, no. 2 (2014): 746–51, https://doi.org/10.1073/pnas.1315088111.

210 **make the community itself the firebreak:** Jack Cohen, "Preventing Residential Fire Disasters During Wildfires," US Department of Agriculture, Forest Service Research, Missoula, MT, n.d., http://www.fria.gr/WARM /chapters/warmCh01Cohen.pdf.

210 **burned the town of Paradise:** Julia Prodis Sulek and Annie Sciacca, "'This Is When I Die' . . . Unforgettable Tales of Escape from the Camp Fire," *San Jose Mercury News,* November 10, 2018, https://www.mercurynews.com/2018/11/10 /this-is-when-i-die-unforgettable-tales-of-escape-from-the-camp-fire/.

210 **most expensive natural disaster:** Petra Löw, "The Natural Disasters of 2018 in Figures," Munich Re, August 1, 2019, https://www.munichre.com /topics-online/en/climate-change-and-natural-disasters/natural-disasters/the -natural-disasters-of-2018-in-figures.html.

210 **economic losses over $1 billion:** National Centers for Environmental Information, "Assessing the U.S. Climate in 2019," January 8, 2020, accessed July 25, 2020, https://www.ncei.noaa.gov/news/national-climate-201912.

211 **abandoned for human habitation:** David Roberts, "3 Key Solutions to California's Wildfire Safety Blackout Mess," Vox, October 22, 2019, https:// www.vox.com/energy-and-environment/2019/10/22/20916820/california-wild fire-climate-change-blackout-insurance-pge.

211 **governor opposed limiting:** Kathleen Ronayne, "California Governor Won't Block Building in High-Fire Areas," AP News, April 16, 2019, https:// apnews.com/b17b5c9200a64466b49f3f605f9202fe.

211 **$1.70 for every dollar:** Dale Kasler, Ryan Sabalow, and Phillip Reese, "'Sticker Shock' for California Wildfire Areas: Insurance Rates Doubled, Poli- cies Dropped," *Sacramento Bee,* July 18, 2019, https://www.sacbee.com/news /politics-government/capitol-alert/article232575652.html.

Chapter 12

213 **closing words from . . . Laurent Fabius:** Kimberly Nicholas (@KA_ Nicholas), "Translator's voice quivering w/ emotion at closing words @Lau- rentFabius: 'The world is holding its breath. It counts on all of us,' #COP21," Twitter, December 12, 2015, 12:13 P.M., https://twitter.com/KA_Nicholas /status/675634614787031040.

213 **a misplaced "shall":** Joshua Keating, "The One Word That Almost Scut- tled the Climate Deal," *Slate,* December 14, 2015, https://slate.com/news -and-politics/2015/12/climate-deal-came-down-to-the-difference-between-shall -and-should.html.

214 **climate citizens:** Kate Knuth, "What It Means to Be a Climate Citizen," Democracy and Climate, September 25, 2019, https://democracyandclimate .com/2019/09/25/what-it-means-to-be-a-climate-citizen/.

215 **entrench further carbon lock-in:** Karen C. Seto et al., "Carbon Lock-In: Types, Causes, and Policy Implications," *Annual Review of Environ- ment and Resources* 41, no. 1 (2016): 425, 445, 446, https://doi.org/10.1146 /annurev-environ-110615-085934.

215 **three political processes:** Steven Bernstein and Matthew Hoffmann, "The Politics of Decarbonization and the Catalytic Impact of Subnational Climate Experiments," *Policy Sciences* (2018): 191, 195, 198, 200, https://doi .org/10.1007/s11077-018-9314-8.

216 **lead the global revenue list:** Felix Todd, "Oil and Gas Companies Earn Most Revenue in Forbes 2019 Largest Firms List," NS Energy, August 20, 2019, https://www.nsenergybusiness.com/news/oil-gas-revenue-forbes -2019/.

216 **five of the ten wealthiest:** "Global 500," *Fortune,* accessed October 22, 2020, https://fortune.com/global500/.

216   **$200 million each year:** InfluenceMap, "Big Oil's Real Agenda on Climate Change" (2019), https://influencemap.org/report/How-Big-Oil -Continues-to-Oppose-the-Paris-Agreement-38212275958aa21196dae3b7622 0bddc.

216   **spent ten times more money:** Robert J. Brulle, "The Climate Lobby: A Sectoral Analysis of Lobbying Spending on Climate Change in the USA, 2000 to 2016," *Climatic Change* 149, nos. 3–4 (2018): 289–303, https://doi .org/10.1007/s10584-018-2241-z.

216   **$60 billion in foregone social benefits:** Kyle C. Meng and Ashwin Rode, "The Social Cost of Lobbying over Climate Policy," *Nature Climate Change* 9, no. 6 (2019): 472–76, https://doi.org/10.1038/s41558-019-0489-6.

216   **"We used to think":** Benjamin Franta, "Facilitating the Green New Deal Through History and Law," presentation at the American Geophysical Union Fall Meeting, December 12, 2019, San Francisco, California.

217   **"political constraints are 'soft'":** J. Jewell and A. Cherp, "On the Politi- cal Feasibility of Climate Change Mitigation Pathways: Is It Too Late to Keep Warming Below 1.5C?" *WIREs Climate Change* (2019): 6–7, https://doi.org /10.1002/wcc.621.

217   **vision for sustainable food:** Johan O. Karlsson et al., "Designing a Fu- ture Food Vision for the Nordics Through a Participatory Modeling Ap- proach," *Agronomy for Sustainable Development* 38, no. 6 (2018), https://doi .org/10.1007/s13593-018-0528-0.

217   **participatory paintings:** Emma L. Johansson and Ellinor Isgren, "Local Perceptions of Land-Use Change: Using Participatory Art to Reveal Direct and Indirect Socioenvironmental Effects of Land Acquisitions in Kilombero Valley, Tanzania," *Ecology and Society* 22, no. 1 (2017), https://doi.org /10.5751/es-08986-220103.

218   **a range of pathways:** Åsa Svenfelt et al., "Scenarios for Sustainable Fu- tures Beyond GDP Growth 2050," *Futures* 111 (2019): 1–14, https://doi.org /10.1016/j.futures.2019.05.001.

218   **top choice was to vote:** M. Hooghe and S. Marien, "How to Reach Mem- bers of Parliament? Citizens and Members of Parliament on the Effectiveness of Political Participation Repertoires," *Parliamentary Affairs* 67, no. 3 (2012): 536–60, https://doi.org/10.1093/pa/gss057.

218   **just one year's worth:** Shaikh M. S. U. Eskander and Sam Fankhauser, "Reduction in Greenhouse Gas Emissions from National Climate Legisla- tion," *Nature Climate Change* 10 (2020): 750–56, https://www.nature.com /articles/s41558-020-0831-z.

219   **electing politicians with good LCV scores:** John Muñoz, Susan Ol- zak, and Sarah A. Soule, "Going Green: Environmental Protest, Policy, and CO_2 Emissions in U.S. States, 1990–2007," *Sociological Forum* 33, no. 2 (2018): 403–21, table 2, https://doi.org/10.1111/socf.12422.

219   **electing more women:** Astghik Mavisakalyan and Yashar Tarverdi, "Gen- der and Climate Change: Do Female Parliamentarians Make Difference?" *European Journal of Political Economy* 56 (2019): 151–64, https://doi.org /10.1016/j.ejpoleco.2018.08.001.

219   **new administrations can undo climate progress:** Nadja Popovich, Livia Albeck-Ripka, and Kendra Pierre-Louis, "The Trump Administration Is Reversing 100 Environmental Rules. Here's the Full List," *The New York*

Times, May 20, 2020, https://www.nytimes.com/interactive/2020/climate/trump-environment-rollbacks.html.

219 **next most effective action:** Hooghe and Marien, "How to Reach Members of Parliament?" tables 3 and 4.

219 **(or climate organization):** Seth Wynes, *SOS: What You Can Do to Reduce Climate Change* (London: Ebury Press, 2019), 73.

220 **phone calls to the local district office:** Daniel Victor, "Here's Why You Should Call, Not Email, Your Legislators," *The New York Times,* November 22, 2016, https://www.nytimes.com/2016/11/22/us/politics/heres-why-you-should-call-not-email-your-legislators.html.

220 **felt very little pressure:** Rebecca Willis, "Building the Political Mandate for Climate Action," Green Alliance (November 2018), https://www.green-alliance.org.uk/resources/Building_a_political_mandate_for_climate_action.pdf.

221 **three-phase strategy to deliver Paris:** Kevin Anderson, "Hope from Despair: Transforming Delusion into Action on Climate Change," Scientists for Global Responsibility Conference, November 26, 2019: 42, 50, https://www.sgr.org.uk/resources/hope-despair-transforming-delusion-action-climate-change.

222 **public needs to understand:** Intergovernmental Panel on Climate Change (IPCC), "Summary for Policymakers," *Global Warming of 1.5°C. An IPCC Special Report on the Impacts of Global Warming of 1.5°C Above Pre-Industrial Levels and Related Global Greenhouse Gas Emission Pathways, in the Context of Strengthening the Global Response to the Threat of Climate Change, Sustainable Development, and Efforts to Eradicate Poverty,* ed. V. Masson-Delmotte et al. (Geneva, Switzerland: IPCC, 2018), D5.6.

223 **people as agents of change:** Karen O'Brien, "Is the 1.5°C Target Possible? Exploring the Three Spheres of Transformation," *Current Opinion in Environmental Sustainability* 31 (2018): 153–60, https://doi.org/10.1016/j.cosust.2018.04.010.

223 **requires the participation:** Elinor Ostrom, *Governing the Commons: The Evolution of Institutions for Collective Action* (Cambridge, UK: Cambridge University Press, 1990).

223 **granting rights of personhood:** Mihnea Tanasescu, "When a River Is a Person: From Ecuador to New Zealand, Nature Gets Its Day in Court," The Conversation, June 19, 2017, https://theconversation.com/when-a-river-is-a-person-from-ecuador-to-new-zealand-nature-gets-its-day-in-court-79278.

223 **Climate Assembly UK:** Daisy Dunne et al., "Q&A: How the 'Climate Assembly' Says the UK Should Reach Net-Zero," Carbon Brief, September 10, 2020, https://www.carbonbrief.org/qa-how-the-climate-assembly-says-the-uk-should-reach-net-zero.

223 **future generations minister:** Oliver Balch, "Meet the World's First 'Minister for Future Generations,'" *The Guardian,* March 2, 2019, https://www.theguardian.com/world/2019/mar/02/meet-the-worlds-first-future-generations-commissioner.

224 **jobs strongly increased support:** Parrish Bergquist, Matto Mildenberger, and Leah C. Stokes, "Combining Climate, Economic, and Social Policy Builds Public Support for Climate Action in the US," *Environmental Research Letters* 15, no. 5 (2020): figure 1, https://doi.org/10.1088/1748-9326/ab81c1.

224 **Arby's restaurants employ more Americans:** Christopher Ingraham, "The Entire Coal Industry Employs Fewer People than Arby's," *The Washington Post,* March 31, 2017, https://www.washingtonpost.com/news/wonk/wp/2017/03/31/8-surprisingly-small-industries-that-employ-more-people-than-coal/.

224 **Ministry of Ecological Transition:** Ellie Anzilotti, "Spain Wants to Phase Out Coal Plants Without Hurting Miners," *Fast Company,* October 26, 2018, https://www.fastcompany.com/90257736/spain-wants-to-phase-out-coal-plants-without-hurting-miners.

224 **Alberta's oil sands are mobilizing:** Chloe Williams, "As Oil Industry Swoons, Tar Sands Workers Look to Renewables for Jobs," *Yale Environment 360,* April 30, 2020, https://e360.yale.edu/features/as-oil-industry-swoons-tar-sands-workers-look-to-renewables-for-jobs.

224 **US public concern for climate:** Jason T. Carmichael and Robert J. Brulle, "Elite Cues, Media Coverage, and Public Concern: An Integrated Path Analysis of Public Opinion on Climate Change, 2001–2013," *Environmental Politics* 26, no. 2 (2016): 247, https://doi.org/10.1080/09644016.2016.1263433.

224 **media often miss opportunities:** Neil Demause, "Media on Climate Crisis: Don't Organize, Mourn," FAIR, January 31, 2020, https://fair.org/home/media-on-climate-crisis-dont-organize-mourn/.

225 **stopped accepting fossil fuel:** Hanna Frick, "*The Guardian* tar efter *Dagens ETC*—stoppar fossila annonser" [*The Guardian* takes after *Dagens ETC*—stops fossil ads], *Dagens Media,* January 29, 2020, https://www.dagensmedia.se/medier/dagspress/the-guardian-tar-efter-dagens-etc-stoppar-fossila-annonser/.

225 ***The Guardian* followed partway:** Jim Waterson, "*Guardian* to Ban Advertising from Fossil Fuel Firms," *The Guardian,* January 29, 2020, https://www.theguardian.com/media/2020/jan/29/guardian-to-ban-advertising-from-fossil-fuel-firms-climate-crisis.

225 **more than thirteen hundred climate-related lawsuits:** Joana Setzer and Rebecca Byrnes, "Global Trends in Climate Change Litigation: 2019 Snapshot," London School of Economics, 2019, 1, 3, http://www.lse.ac.uk/GranthamInstitute/wp-content/uploads/2019/07/GRI_Global-trends-in-climate-change-litigation-2019-snapshot-2.pdf.

225 **emissions cuts of 25 percent:** Isabella Kaminski, "Historic Urgenda Climate Ruling Upheld by Dutch Supreme Court," The Climate Docket, December 20, 2019, https://www.climatedocket.com/2019/12/20/urgenda-climate-ruling-netherlands-supreme-court/.

225 **Urgenda won . . . close coal-fired power:** Tonya Mosley, "Dutch Court Says Government Inaction on Climate Change Violates Human Rights," WBUR, January 13, 2020, https://www.wbur.org/hereandnow/2020/01/13/netherlands-climate-change-human-rights.

225 **Sophie Marjanac argued:** Sophie Marjanac, "Attributing Climate Impacts to Major Fossil Fuel Companies," side event at the 23rd Conference of the Parties to the United Nations Framework Convention on Climate Change (COP23), Bonn, Germany, November 16, 2017.

226 **"Our citizens keep marching":** David Hudson, "President Obama: 'No Nation Is Immune' to Climate Change," White House, September 23, 2014, https://obamawhitehouse.archives.gov/blog/2014/09/23/president-obama-no-nation-immune-climate-change.

226 **against the wishes of almost 70 percent:** Ed Maibach, Anthony Leiserowitz, and Jennifer Marlon, "Should the US Stay in the Paris Agreement? A Majority of Democrats and Republicans Think So," The Conversation, May 15, 2017, https://theconversation.com/should-the-us-stay-in-the-paris-agreement-a-majority-of-democrats-and-republicans-think-so-77455.

227 **footnoted protest sign:** For the story, see Kimberly Nicholas, "A Climate Change Curriculum Based on Synthesis Science," Kimberly Nicholas, http://www.kimnicholas.com/climate-change-curriculum.html.

227 **more than 6 million people:** Matthew Taylor, Jonathan Watts, and John Bartlett, "Climate Crisis: 6 Million People Join Latest Wave of Global Protests," *The Guardian,* September 27, 2019, https://www.theguardian.com/environment/2019/sep/27/climate-crisis-6-million-people-join-latest-wave-of-worldwide-protests.

227 **an exponential rise:** Elizabeth Sawin (@bethsawin), "Small changes can snowball. One person can reach out to one more. Courage is contagious. Concern spreads when validated. Organizing works. And a movement whose rate of growth is not constant but is itself growing? That is a force for transformation," Twitter, September 22, 2019, 8:15 P.M., https://twitter.com/bethsawin/status/1175836141167218688.

228 **"not be an individual":** Ken Jones, "Bill McKibben quote, 'Don't be an individual,'" 2015, video, 0:16, https://www.youtube.com/watch?v=m_-ts DIoKBw.

228 **adopted a climate framework:** The Swedish "Climate Framework for Higher Education Institutions—Guidelines" is a good template for assessing high-impact areas and identifying key steps for change. See "The Climate Framework," KTH Royal Institute of Technology, https://www.kth.se/en/om/miljo-hallbar-utveckling/klimatramverket-1.903489.

229 **25 percent of a population:** Damon Centola et al., "Experimental Evidence for Tipping Points in Social Convention," *Science* 360, no. 6393 (2018): 1116–19, https://doi.org/10.1126/science.aas8827.

229 **can catalyze rapid social change:** I. M. Otto et al., "Social Tipping Dynamics for Stabilizing Earth's Climate by 2050," *Proceedings of the National Academy of Sciences of the United States of America* 117, no. 5 (February 4, 2020): 2354–65, https://doi.org/10.1073/pnas.1900577117.

229 **Social movements and political campaigns:** John Muñoz, Susan Olzak, and Sarah A. Soule, "Going Green: Environmental Protest, Policy, and CO_2 Emissions in U.S. States, 1990–2007," *Sociological Forum* 33, no. 2 (2018): 403–21, https://doi.org/10.1111/socf.12422.

230 **record-breaking crowd:** Adam Westin, "Rekordstor klimatmarsch i Stockholm" [Record-breaking climate march in Stockholm], *Aftonbladet,* September 27, 2019, https://www.aftonbladet.se/nyheter/a/LALwBV/rekordstor-klimatmarsch-i-stockholm.

Chapter 13

233 **positive social tipping point:** I. M. Otto et al., "Social Tipping Dynamics for Stabilizing Earth's Climate by 2050," *Proceedings of the National Academy of Sciences of the United States of America* 117, no. 5 (2020): 2354–65, https://doi.org/10.1073/pnas.1900577117.

233 **catastrophic ecological tipping points:** Timothy M. Lenton et al.,
     "Climate Tipping Points—Too Risky to Bet Against," *Nature* 575 (2019): 592–95,
     https://www.nature.com/articles/d41586-019-03595-0#correction-0.
235 **finding bright spots:** Chip Heath and Dan Heath, *Switch: How to Change
     Things When Change Is Hard* (Waterville, ME: Thorndike Press, 2011).
236 **being a good ancestor requires:** Bina Venkataraman, "Why You Should
     Think About Being a Good Ancestor—and 3 Ways to Start Doing It," TED
     video, August 17, 2019, https://ideas.ted.com/why-you-should-think-about
     -being-a-good-ancestor-and-3-ways-to-start-doing-it/.
236 **present bias favoring short-term gratification:** Ben Yagoda, "The
     Cognitive Biases Tricking Your Brain," *The Atlantic,* September 2018, https://
     www.theatlantic.com/magazine/archive/2018/09/cognitive-bias/565775/.
236 **"estranged" . . . "aged avatars":** H. E. Hershfield et al., "Increasing
     Saving Behavior Through Age-Progressed Renderings of the Future Self,"
     Journal of Marketing Research 48 (2011): 3, 8, https://doi.org/10.1509/jmkr.48
     .SPL.S23.
236 **activists from Sunrise Movement:** Sunrise Movement, "Climate Leg-
     acy Time Capsule," https://www.sunrisemovement.org/time-capsule.
237 **Future Library, which is collecting:** Katie Paterson, Future Library,
     https://www.futurelibrary.no/. This paragraph was adapted from a review I
     wrote for The Reading Lists, "The Best Books on Climate Change," 2018,
     https://www.thereadinglists.com/best-books-on-climate-change/.
238 **increased their parents' concern:** Danielle F. Lawson et al., "Children
     Can Foster Climate Change Concern Among Their Parents," *Nature Climate
     Change* 9, no. 6 (2019): 458–62, https://doi.org/10.1038/s41558-019-0463-3.
238 **eleven-year-old Lilly:** Lillys Plastic Pickup (@lillyspickup), "#ICanVote
     at the #EUelections2019 thanks to my grandpa who will vote for me. We want
     #ClimateAction. Talk to your (grand) parents and ask for their vote too. Let's
     make politicians listen. #ClimateCoup #ParentsForClimate YouthForCli-
     mate @Europarl_EN," Twitter, April 10, 2019, 3:29 P.M., https://twitter.com
     /lillyspickup/status/1115970080062484482.
239 **Their ask is that:** Greta Thunberg, *No One Is Too Small to Make a Differ-
     ence* (New York: Penguin Books, 2019): 32–33.

Conclusion

243 **about a quarter:** "Global Carbon Budget," Global Carbon Project, data
     downloaded September 11, 2020, https://www.globalcarbonproject.org/car
     bonbudget/19/visualisations.htm.
244 **the end of history illusion:** Jordi Quoidbach, Daniel T. Gilbert, and
     Timothy D. Wilson, "The End of History Illusion," *Science* 339, no. 6115
     (2013): 96–98, https://doi.org/10.1126/science.1229294.

Index

About the Author

Simon Charles Florian Rose

Dr. Kimberly Nicholas has published more than fifty articles on climate and sustainability in leading peer-reviewed journals, and her research has been featured in outlets including *The New York Times*, *The Washington Post*, *The Atlantic*, *USA Today*, *BuzzFeed*, and more. She has also been profiled in *Elle* and *The Guardian*, and gives appearances at around fifty lectures each year, such as the recent Climate Change Leadership summit in Porto. Dr. Nicholas is associate professor of Sustainability Science at Lund, the highest-ranked university in Sweden. Born and raised on her family's vineyard in Sonoma, California, she studied the effect of climate change on the California wine industry for her PhD in Environment and Resources at Stanford University.

CONNECT ONLINE

www.kimnicholas.com

🐦 KA_Nicholas

📷 kimberlynicholasphd

📘 Kimberly.Nicholas.52